D0712570

de Gruyter Series in Nonlinear Analysis and Applications 9

Carlo Bardaro
Julian Musielak
Gianluca Vinti

Nonlinear Integral Operators and Applications

Walter de Gruyter · Berlin · New York 2003

Authors

Carlo Bardaro
Dipartimento di Matematica
e Informatica
Via Vanvitelli, 1
06123 PERUGIA
ITALY

Julian Musielak
Krasinskiego, 8D
60869 POZNAN
POLAND

Gianluca Vinti
Dipartimento di Matematica
e Informatica
Via Vanvitelli, 1
06123 PERUGIA
ITALY

Mathematics Subject Classification 2000: *Primary:* 41-02; 26A45, 26A46, 40F05, 41A25, 41A35, 41A46, 41A65, 45G10, 45N05, 45P05, 46E30, 47B38, 47J05, 94A20; *secondary:* 26A15, 26A16, 26A51, 26D10, 26D15, 40C15, 44A35, 94A08, 94A12.

Keywords: Kernel functional, convolution type operators, singularity, rate of approximation, modular spaces, modulus of continuity, functions of bounded variation, nonlinear integral equations, sampling type operators, summability.

⊗ Printed on acid-free paper which falls within the guidelines of the ANSI to ensure permanence and durability.

Library of Congress Cataloging-in-Publication Data

Bardaro, Carlo.
Nonlinear integral operators and applications / Carlo Bardaro, Julian Musielak, Gianluca Vinti.
p. cm. – (De Gruyter series in nonlinear analysis and applications, ISSN 0941-813X ; 9)
Includes bibliographical references and index.
ISBN 3-11-017551-7 (alk. paper)
1. Integral operators. 2. Nonlinear operators. I. Musielak, Julian, 1928– II. Vinti, Gianluca. III. Title. IV. Series.
QA329.6.B37 2003
515′.723–dc21 2003042878

ISBN 3-11-017551-7

Bibliographic information published by Die Deutsche Bibliothek

Die Deutsche Bibliothek lists this publication in the Deutsche Nationalbibliografie; detailed bibliographic data is available in the Internet at <http://dnb.ddb.de>.

Printed in Germany.
Cover design: Thomas Bonnie, Hamburg
Typeset using the authors' TEX files: I. Zimmermann, Freiburg
Printing and binding: Hubert & Co. GmbH & Co. Kg, Göttingen

To our parents, for their profound influence on our lives

Preface

One hundred years ago I. Fredholm published in 1903 his famous paper [103] on linear integral equations. Since then, linear integral operators have become important tools in many areas, including the theory of Fourier series and Fourier integrals, approximation theory and summability theory and the theory and practice of solving integral and differential equations. In the case of integral and differential equations, applications were soon extended beyond the confines of linear operators. In approximation theory, however, applications were limited to linear operators because the notion of singularity of an integral operator was closely connected with its linearity. Then, about twenty years ago, the concept of singularity was extended to cover the case of nonlinear integral operators [152]. Since that time a number of papers have appeared that are devoted to the investigation of the role played by nonlinear integral operators in approximation theory and related subjects. For example, the study of certain discrete operators, the so-called "generalized sampling operators", provides the basis for several applications in signal analysis. From this work a new theory of signal processing in the nonlinear setting may be developed. This is of considerable interest, not only from the mathematical point of view, but also for applications in engineering. For example, the reconstruction of signals by means of nonlinear sampling-type operators may describe nonlinear models that are suitable for the processing of some class of signals.

Recently, a number of important contributions by P. L. Butzer and his school have been made to exponential sampling and Mellin–Fourier approximation theory. As a consequence, it seems very useful for us to have at our disposal a nonlinear version of the Mellin convolution operators with associated approximation properties in various function spaces.

The purpose of this book is to present the fundamental theoretical results along with a variety of recent applications. We consider nonlinear integral operators, replacing linearity by generalized Lipschitz conditions for kernel functions generating the operators and satisfying suitable singularity assumptions. Applications in approximation theory and summability theory require a notion of convergence for sequences or directed families of such operators. We replace the standard setting of normed linear spaces by the more general one of modular linear spaces. This extends the field of applications and enables us to give a unitary approach to various kinds of approximation problems. For example, classical approximation theorems for linear or nonlinear integral operators in L^p-spaces, in Orlicz spaces and in other functional spaces can be derived by a unique method.

The prerequisites needed to study this book consist of the theory of measure and integral and some fundamental knowledge of functional analysis. With the exception of the above, the material in the book is selfcontained.

Chapter 1 is of a preliminary character and contains material concerning kernel functions generating the operators and an introduction to the theory of modular spaces. In this chapter we also discuss some special conditions concerning modulars, needed for our purposes.

A fundamental role in application of integral operators in approximation theory is played by the notion of modulus of continuity of functions in a modular space, which is considered in Chapter 2. A convergence theorem for such moduli requires the notion of absolute continuity of the modular, which is applied through a Lebesgue-type dominated convergence theorem in the respective modular spaces.

In Chapter 3 we describe applications of nonlinear integral operators to approximation theory, presenting an embedding theorem, an estimate of the error of modular approximation and a theorem concerning modular convergence to zero of the error in the case of a family of operators. The rates of modular approximation in modular Lipschitz classes are also estimated. Finally, we present results in the case of a nonlinear integral operator being split into a linear part and a nonlinear perturbation.

The above considerations are continued in Chapter 4, where Urysohn's integral operators with homogeneous kernels are investigated. We give again an estimation of the error and a result on convergence of the error in the sense of modular convergence. An application is given to nonlinear weighted Mellin convolution operators.

Chapter 5 contains results concerning conservative nonlinear summability methods defined by families of nonlinear integral operators.

Prior to Chapter 6 we assume in most of the theorems that the modular which generates the convergence is monotone. This holds for example in the case of Orlicz spaces and a number of their generalizations. In Chapter 6 we consider modulars generating the space of functions of generalized bounded variation, which are not monotone. We obtain embedding-type inequalities and we also consider the case of superposition of nonlinear integral operators. Convergence in generalized variation is also studied, for a special class of nonlinear operators, namely the nonlinear Mellin-type convolution operators. The problem of convergence for general nonlinear integral operators is still an open problem.

In Chapter 7 a solution is given to the problem of existence for the domain of a nonlinear convolution-type integral operator. We also present some results concerning existence of solutions of the respective nonlinear integral equations through the fixed point principles of Banach and Schauder.

Chapter 8 is devoted to uniform approximation of continuous functions by means of nonlinear sampling type operators. The chapter begins with an introduction to the theory of sampling. We give embedding results, estimation of the error of approximation and a convergence theorem. Moreover, rates of uniform approximation are considered. An application to regular methods of summability is also given.

In Chapter 9, results of Chapter 8 are extended from uniform approximation to the case of modular approximation. We consider problems of modular convergence and modular approximation for nonlinear sampling type operators, with special emphasis to the case of generalized Orlicz type modulars. We also give an application to regular methods of summability.

This book is a result of collaboration between its authors during the years 1993–2002. Such collaboration would not have been possible without grants from the Consiglio Nazionale delle Ricerche (CNR) in Italy. This support enabled the second author to spend some time in Perugia each year, and he wishes to express his gratitude for this generosity. The authors wish to acknowledge the hospitality of the University of Perugia and would like to express their gratitude to the Faculty of Mathematics and Computer Science of the A. Mickiewicz University in Poznań for the grant GN-11/99 supporting the contacts between them over the period from 1998 to 2001.

The authors also wish to thank many colleagues for helpful discussions on the material presented here. In addition, it gives them special pleasure to express their gratitude to Professors P. L. Butzer, R. J. Nessel and R. L. Stens of the RWTH Aachen, who read the text and made valuable suggestions, and to Professor A. M. Arthurs of the University of York for his careful revision of the language of the text.

Perugia and Poznań, April 2003

Carlo Bardaro
Julian Musielak
Gianluca Vinti

Contents

Chapter 1

Kernel functionals and modular spaces

1.1 Kernel functionals

Let (Ω, Σ, μ) be a measure space with a σ-finite, complete measure μ. Let $L^0(\Omega)$ denote the space of all extended real-valued, Σ-measurable functions on Ω, finite a.e. (μ-almost everywhere), with equality μ-a.e. A functional $K : \Omega \times \Omega \times \operatorname{Dom} K \to \mathbb{R}$, where $\operatorname{Dom} K \subset L^0(\Omega)$, will be called a *kernel functional*, if for every $f \in \operatorname{Dom} K$, the functional $K(s, t, f)$ is measurable in the product $\Omega \times \Omega$, and if $K(s, t, 0) = 0$, for all $s, t \in \Omega$. The set $\operatorname{Dom} K$ is called the *domain of the kernel functional* K.

Example 1.1. Let $K_1 : \Omega \times \Omega \times \mathbb{R} \to \mathbb{R}$ be such that $K_1(s, t, u)$ is measurable as a function of $(s, t) \in \Omega \times \Omega$, for every $u \in \mathbb{R}$, and is continuous as a function of $u \in \mathbb{R}$ for all $(s, t) \in \Omega \times \Omega$. Let us suppose that $K_1(s, t, 0) = 0$ for all $s, t \in \Omega$. Obviously K_1 is a kernel functional in the domain $\operatorname{Dom} K_1$, consisting of the set of all constant functions in $L^0(\Omega)$. One may define another kernel functional K by means of K_1, taking $K(s, t, f) = K_1(s, t, f(t))$ for $s, t \in \Omega$. Obviously $\operatorname{Dom} K = L^0(\Omega)$.

Example 1.2. Let $K_2 : \Omega \times \mathbb{R} \to \mathbb{R}$ be such that $K_2(t, u)$ is Σ-measurable in Ω for every $u \in \mathbb{R}$ and is continuous in $\mathbb{R}$ for all $t \in \Omega$ (i.e. K_2 is a Carathéodory function). Let $K_2(t, 0) = 0$ for all $t \in \Omega$. Moreover let us suppose that there is an operation $+ : \Omega \times \Omega \to \Omega$, which is measurable as a function from the product $\Omega \times \Omega$ to Ω. Then $K(s, t, f) = K_2(t, f(t + s))$ is a kernel functional and $\operatorname{Dom} K$ contains all constant functions. It is easily seen that if $\Omega = \mathbb{R}^N$ and $t + s$ means the sum of vectors t and s, μ is the Lebesgue measure in the σ-algebra of Lebesgue measurable sets in $\mathbb{R}^N$, then $\operatorname{Dom} K = L^0(\Omega)$.

The functions K_1, K_2 of the previous examples, with the above assumptions, will be called *kernel functions*.

Example 1.3. Let $p : \Omega \to \mathbb{R}$ be Σ-measurable and let $K(s, t, f) = K_3(s, t, f) := p(t) f(s)$ for every $f \in L^0(\Omega)$. Obviously K is a kernel functional with domain $\operatorname{Dom} K = L^0(\Omega)$.

Since from the measurability of a kernel functional $K(s, t, f)$ in $\Omega \times \Omega$ there follows its Σ-measurability as a function of the variable t for μ-a.e. $s \in \Omega$, so the kernel functional generates an *integral operator* T by the formula:

$$(Tf)(s) = \int_\Omega K(s, t, f)\, d\mu(t), \tag{1.1}$$

in the case the integral at the right-hand side of (1.1) exists for μ-a.e. $s \in \Omega$ and is a Σ-measurable function of $s \in \Omega$. The set of all such functions f is called the *domain of the operator* T and will be denoted by Dom T.

Example 1.4. Let us examine the integral operator T of a kernel $K(s, t, f) = K_2(t, f(t + s))$ (Example 1.2) in the special case when $(\Omega, +)$ is a unimodular, locally compact Hausdorff topological group with Haar measure μ. Then

$$(Tf)(s) = \int_\Omega K_2(t, f(t+s))\, d\mu(t) = \int_\Omega K_2(t - s, f(t))\, d\mu(t).$$

Denoting $\check{g}(t, u) = g(-t, u)$ for any $g : \Omega \times \mathbb{R} \to \mathbb{R}$, $t \in \Omega$, $u \in \mathbb{R}$, we have

$$(Tf)(s) = \int_\Omega \check{K}_2(s - t, f(t))\, d\mu(t). \tag{1.2}$$

In case when $\check{K}_2(t, u) = K(t)u$ for $t \in \Omega$, $u \in \mathbb{R}$ and $K : \Omega \to \mathbb{R}$, the integral (1.2) becomes a convolution: $Tf = K * f$.

In the general case we call the operator T defined by the formula (1.1) with $K(s, t, f) = K_2(t, f(t + s))$ a *convolution-type operator*. Later we consider such operators without group structure of $(\Omega, +)$.

Let us still remark that the kernel functional written in the form used in the formula (1.2) is a special case of the kernel functional $K(s, t, f) = K_1(s, t, f(t))$ from Example 1.1 if we put $K_1(s, t, u) = K_2(t - s, u) = \check{K}_2(s - t, u)$.

The operator (1.1) in the case of $K(s, t, f) = K_1(s, t, f(t))$ (Example 1.2) is known as the *Urysohn operator*. A special case of this operator, called the *Hammerstein operator* is obtained by taking $K_1(s, t, u) = K^{(1)}(s, t)K^{(2)}(t, u)$. Both operators are studied mainly in connection with the theory of nonlinear integral equations of the form

$$\int_\Omega K_1(s, t, f(t))\, d\mu(t) = f(s) + g(s)$$

and

$$\int_\Omega K^{(1)}(s, t)K^{(2)}(t, f(t))\, d\mu(t) = f(s) + g(s)$$

with known g and unknown f, called the *Urysohn integral equation* and *Hammerstein integral equation*, respectively. References to the above operators and integral equations may be found e.g. in [133], [132], [131], [113], [182], [193], [194], [195]. We shall consider some nonlinear integral equations of the above type in Chapter 7.

Finally, let us remark that in the case of the kernel functional $K(s, t, f) = p(t)f(s)$ from Example 1.3, the operator given in (1.1) is of the form

$$(Tf)(s) = f(s) \int_\Omega p(t)\, d\mu(t).$$

Supposing further that $p \in L^1(\Omega)$ and writing $\lambda = \int_\Omega p(t)\, d\mu(t)$, we obtain Dom $T = L^0(\Omega)$ and $(Tf)(s) = \lambda f(s)$, for $s \in \Omega.xp$

1.2 Modular spaces and modular convergence

In order to investigate any kind of convergence process for sequences or families of integral operators of the form (1.1), one has to specify some function spaces which are subspaces of $L^0(\Omega)$ and to provide these subspaces with a suitable notion of convergence. This can be done by taking normed linear spaces contained algebraically in $L^0(\Omega)$. However, in order to obtain results on a level of generality allowing a wide spectrum of applications it is more suitable to replace the notions of a norm and a normed linear space by those of a modular and a modular space. Here, we shall limit ourselves to definitions in the case of function spaces only. A more general treatment may be found e.g. in [153].

Let $\mathcal{X}$ be a linear space of real-valued functions (extended real-valued, eventually), defined on a nonempty set Ω, with equality everywhere (or almost everywhere). A functional $\rho : \mathcal{X} \to \widetilde{\mathbb{R}}_0^+ = [0, +\infty]$ is called a *modular* on $\mathcal{X}$, if it satisfies the following conditions for arbitrary $f, g \in \mathcal{X}$:

(1) $\rho(f) = 0$ if and only if $f = 0$,

(2) $\rho(-f) = \rho(f)$,

(3) $\rho(\alpha f + \beta g) \leq \rho(f) + \rho(g)$, for $\alpha, \beta \geq 0, \alpha + \beta = 1$.

The modular ρ is called a *convex modular* if (3) is replaced by

(3)′ $\rho(\alpha f + \beta g) \leq \alpha\rho(f) + \beta\rho(g)$, for $\alpha, \beta \geq 0, \alpha + \beta = 1$.

It is easily seen that conditions (3) and (3)′ may be extended by induction to any finite number of terms, i.e. (3) is equivalent to

$$\rho\Big(\sum_{j=1}^{n} \alpha_j f_j\Big) \leq \sum_{j=1}^{n} \rho(f_j),$$

for $\alpha_j \geq 0$, $\sum_{j=1}^{n} \alpha_j \leq 1$ and (3)′ is equivalent to

$$\rho\Big(\sum_{j=1}^{n} \alpha_j f_j\Big) \leq \sum_{j=1}^{n} \alpha_j \rho(f_j),$$

for $\alpha_j \geq 0$, $\sum_{j=1}^n \alpha_j \leq 1$, for arbitrary $f_1, f_2, \dots, f_n \in \mathcal{X}$.

Moreover, for any $f \in \mathcal{X}$ the function $\rho(\alpha f)$ of the variable $\alpha > 0$, is nondecreasing.

The modular ρ is called *monotone*, if $f, g \in \mathcal{X}$ and $|f| \leq |g|$ imply $\rho(f) \leq \rho(g)$; if ρ is monotone, and $f \in \mathcal{X}$ implies $|f| \in \mathcal{X}$, then $\rho(f) = \rho(|f|)$ for every $f \in \mathcal{X}$. Indeed, taking in the above definition $g = |f|$, we obtain $\rho(f) \leq \rho(|f|)$, and taking $f = |g|$, we get $\rho(|g|) \leq \rho(g)$ for any $f, g \in \mathcal{X}$.

If ρ is a modular on $\mathcal{X}$, then the *modular space $\mathcal{X}_\rho$ generated by the modular ρ* is defined as

$$\mathcal{X}_\rho = \{f \in \mathcal{X} : \rho(\lambda f) \to 0, \text{ as } \lambda \to 0^+\}.$$

It is easily verified that if ρ is convex, then $\mathcal{X}_\rho$ is the set of functions $f \in \mathcal{X}$ for which $\rho(\lambda_0 f) < +\infty$ for some $\lambda_0 > 0$. It is clear that $\mathcal{X}_\rho$ is a linear subspace of the space $\mathcal{X}$. There holds (see [153])

Theorem 1.1. (a) *If ρ is a modular on $\mathcal{X}$, then $\mathcal{X}_\rho$ is an F-normed linear space with F-norm $|||\cdot|||_\rho$ defined by*

$$|||f|||_\rho = \inf\left\{u > 0 : \rho\left(\frac{f}{u}\right) \leq u\right\}, \quad \textit{for } f \in \mathcal{X}_\rho.$$

(b) *If ρ is a convex modular on $\mathcal{X}$, then $\mathcal{X}_\rho$ is a normed linear space with norm $\|\cdot\|_\rho$ defined by*

$$\|f\|_\rho = \inf\left\{u > 0 : \rho\left(\frac{f}{u}\right) \leq 1\right\}, \quad \textit{for } f \in \mathcal{X}_\rho.$$

(c) *If $|||f|||_\rho < 1$, (resp. $\|f\|_\rho < 1$), then $\rho(f) \leq |||f|||_\rho$, (resp. $\rho(f) \leq \|f\|_\rho$).*

Let us recall that a norm $\|\cdot\|$ is a nonnegative functional on a real linear space $\mathcal{X}$, such that $\|f\| = 0$ if and only if $f = 0$, $\|f + g\| \leq \|f\| + \|g\|$ (triangle inequality) and $\|cf\| = |c|\|f\|$ (homogeneity), for any $f, g \in \mathcal{X}$, $c \in \mathbb{R}$.

An F-norm $|||\cdot|||$ is a nonnegative functional on a real linear space $\mathcal{X}$, such that $|||f||| = 0$ if and only if $f = 0$, $|||f + g||| \leq |||f||| + |||g|||$, and the conditions $c_n \to c$, $|||f_n - f||| \to 0$, imply $|||c_n f_n - cf||| \to 0$ as $n \to +\infty$, for any $f, g, f_n \in \mathcal{X}$, $c_n, c \in \mathbb{R}$.

The couple $(\mathcal{X}, \|\cdot\|)$ is called a normed linear space, and the couple $(\mathcal{X}, |||\cdot|||)$ is called an F-normed linear space. A normed linear space is always F-normed but not conversely.

Proof of Theorem 1.1. We limit ourselves to the case (b), leaving the case (a) to the reader. If $f \in \mathcal{X}_\rho$, then $\rho(f/n) \to 0$, as $n \to +\infty$. Hence the set $\{u > 0 : \rho(f/u) \leq 1\}$ is nonempty; in fact it is a halfline. If $\|f\|_\rho = 0$, then this halfline starts at $u = 0$, whence $\rho(f/u) \leq 1$ for all $u > 0$. By convexity of ρ, we have for $0 < u \leq 1$,

$$\rho(f) = \rho\left(u\frac{f}{u}\right) \leq u\rho(f/u) \leq u.$$

Taking $u \to 0^+$ we get $\rho(f) = 0$ and consequently $f = 0$. In order to get the triangle inequality, let us take any $\varepsilon > 0$ and let us put $u = \|f\|_\rho + \varepsilon$, $v = \|g\|_\rho + \varepsilon$, where $f, g \in \mathcal{X}_\rho$. Then $\rho(f/u) \leq 1$ and $\rho(g/u) \leq 1$. By convexity of ρ, we obtain

$$\rho\left(\frac{f+g}{u+v}\right) = \rho\left(\frac{u}{u+v}\frac{f}{u} + \frac{v}{u+v}\frac{g}{v}\right)$$
$$\leq \frac{u}{u+v}\rho(f/u) + \frac{v}{u+v}\rho(g/u) \leq \frac{u}{u+v} + \frac{v}{u+v} = 1.$$

Thus $\|f+g\|_\rho \leq u + v = \|f\|_\rho + \|g\|_\rho + 2\varepsilon$. Since $\varepsilon > 0$ is arbitrary, we obtain the triangle inequality for $\|\cdot\|_\rho$. Finally, we have for $f \in \mathcal{X}_\rho$ and $c \in \mathbb{R}$

$$\|cf\|_\rho = \inf\left\{u > 0 : \rho\left(\frac{|c|f}{u}\right) \leq 1\right\} = |c| \inf\left\{\frac{u}{|c|} > 0 : \rho\left(\frac{f}{(u/|c|)}\right) \leq 1\right\}$$
$$= |c|\|f\|_\rho.$$

We shall prove (c) in the case of the norm $\|\cdot\|_\rho$. We may suppose that $0 < \|f\|_\rho < 1$. By convexity of ρ we obtain for any $\varepsilon > 0$ such that $\|f\|_\rho + \varepsilon < 1$

$$\rho(f) = \rho\left[(\|f\|_\rho + \varepsilon)\frac{f}{\|f\|_\rho + \varepsilon}\right] \leq (\|f\|_\rho + \varepsilon)\rho\left[\frac{f}{\|f\|_\rho + \varepsilon}\right] \leq \|f\|_\rho + \varepsilon,$$

and taking $\varepsilon \to 0^+$ we obtain $\rho(f) \leq \|f\|_\rho$.

The proofs in the case of the F-norm $|||\cdot|||_\rho$ are analogous to the above ones. □

Example 1.5. (a) If $(\mathcal{X}, \|\cdot\|)$ is a normed linear space, then the functional $\rho(\cdot) = \|\cdot\|$ is a convex modular in $\mathcal{X}$, as follows from the definition of ρ, immediately. Moreover, we have

$$\mathcal{X}_\rho = \{f \in \mathcal{X} : \|\lambda f\| \to 0 \text{ as } \lambda \to 0^+\} = \mathcal{X}$$

and

$$\|f\|_\rho = \inf\{u > 0 : \|f/u\| \leq 1\} = \|f\|,$$

for every $f \in \mathcal{X}$. This shows that the notions of a convex modular and of a modular space generalize those of a norm and a normed linear space.

(b) Let (Ω, Σ, μ) be a measure space with a σ-finite, complete measure μ. Let $\varphi : \mathbb{R}_0^+ \to \mathbb{R}_0^+$ be a nondecreasing (resp. a convex) continuous function with $\varphi(0) = 0$, $\varphi(0) > 0$, for $u > 0$, $\varphi(u) \to +\infty$ as $u \to +\infty$; such a function will be called a *φ-function* (resp. a *convex φ-function*). Then it is easily shown that

$$\rho(f) = I_\varphi(f) = \int_\Omega \varphi(|f(t)|)\, d\mu(t) \tag{1.3}$$

is a modular (resp. a convex modular) on the space $L^0(\Omega)$. We call I_φ an *Orlicz modular* in $L^0(\Omega)$. The respective modular space $L^0_\rho(\Omega)$ is called an *Orlicz space* and it is denoted by $L^{\varphi}(\Omega, \Sigma, \mu)$, or briefly by $L^{\varphi}(\Omega)$. If $\Omega = \mathbb{N} = \{1, 2, \dots\}$ and μ is the counting measure in Ω, the respective Orlicz space is denoted by ℓ^{φ} and it is called the *sequential Orlicz space* .

If $\varphi(u) = u^p$ for $u \geq 0$, $p \geq 1$, then $L^{\varphi}(\Omega) = L^p(\Omega)$ and the norm $\|\cdot\|_{I_\varphi}$ in $L^0(\Omega)$ is equal to

$$\begin{aligned} \|f\|_{I_\varphi} &= \inf\left\{u > 0 : \int_\Omega \left|\frac{f(t)}{u}\right|^p d\mu(t) \leq 1\right\} \\ &= \left(\int_\Omega |f(t)|^p \, d\mu(t)\right)^{1/p} = \|f\|_{L^p(\Omega)}. \end{aligned}$$

(c) Let (Ω, Σ, μ) be as in the case (b). Let $\varphi : \Omega \times \mathbb{R}_0^+ \to \mathbb{R}_0^+$ be such that $\varphi(\cdot, u)$ is Σ-measurable for each $u \geq 0$ and $\varphi(t, :)$ is a φ-function (resp. a convex φ-function) for every $t \in \Omega$. Let

$$\rho(f) = I_\varphi(f) = \int_\Omega \varphi(t, |f(t)|) \, d\mu(t). \tag{1.4}$$

Then ρ is a modular (resp. a convex modular) in $L^0(\Omega)$. The modular space $L^0_\rho(\Omega)$ generated by $\rho = I_\varphi$ is called a *generalized Orlicz space* or a *Musielak–Orlicz space* and it is denoted by $L^{\varphi}(\Omega, \Sigma, \mu)$, or briefly by $L^{\varphi}(\Omega)$. If $\varphi(t, u)$ is independent of the variable t, the Musielak–Orlicz space is reduced to the Orlicz space. Another special case is provided by $\varphi(t, u) = |u|^{p(t)}$, where p is Σ-measurable and $p(t) \geq 1$, for $t \in \Omega$. The respective modular I_φ is convex and the modular space $L^0_\rho(\Omega)$ is equal to the space $L^{p(t)}(\Omega)$ of Σ-measurable functions f, integrable with variable powers $p(t)$.

(d) The following generalization of the case (c) is obtained, assuming (μ_n) to be a sequence of σ-finite measures in Σ, absolutely continuous with respect to μ. Then, denoting by $a_n(t)$, $(t \in \Omega)$, the Radon–Nikodym derivatives of μ_n, we write

$$\rho(f) = \sup_n \int_\Omega a_n(t)\varphi(t, |f(t)|) \, d\mu(t).$$

This is again a modular (a convex modular if φ is a convex φ- function depending on a parameter) in $L^0(\Omega)$. In case of the counting measure $\mu = \mu_{\mathbb{N}}$ in the set $\mathbb{N} = \{1, 2, \dots\}$ of natural numbers, this modular has the form

$$\rho(x) = \sup_n \sum_{j=1}^{\infty} a_{nj}\varphi_j(t_j),$$

and it generates the modular space $\mathcal{X}_\rho$ of real sequences $x = (t_j)$ such that $\rho(\lambda x) \to 0$ as $\lambda \to 0^+$. Sequences $x \in \mathcal{X}_\rho$ such that

$$\lim_{n \to +\infty} \sum_{j=1}^{\infty} a_{nj}\varphi_j(t_j) = 0$$

are called *strongly* (A, φ)*-summable to* 0, where $A = (a_{nj})$ and $\varphi = (\varphi_j)$. The space $T_0(A, \varphi)$ of such sequences forms a subspace of the modular space $\mathcal{X}_\rho$. The reader is encouraged to formulate analogous notions in case of a general measure space in place of $(\mathbb{N}, 2^{\mathbb{N}}, \mu_{\mathbb{N}})$.

(e) Let m be a measure on an interval $[a, b[\subset \mathbb{R}$, where b may be $+\infty$, defined on the σ-algebra of all Lebesgue measurable subsets of $[a, b[$. Let W be a nonempty set of indices and let $(a_w(\cdot))_{w\in W}$ be a family of Lebesgue measurable positive real-valued functions on $[a, b[$. Moreover, let $\Phi : [a, b[\times\mathbb{R}_0^+ \to \mathbb{R}_0^+$ be a function satisfying the following conditions:

1) $\Phi(x, u)$ is a nondecreasing, continuous function of $u \geq 0$, for every $x \in [a, b[$,

2) $\Phi(x, 0) = 0$, $\Phi(x, u) > 0$ for $u > 0$, and $\Phi(x, u) \to +\infty$ as $u \to +\infty$, for every $x \in [a, b[$,

3) there exists $\lim_{x\to b^-} \Phi(x, u) = \widetilde{\Phi}(u) < +\infty$ for every $u \geq 0$,

4) $\Phi(x, u)$ is a Lebesgue measurable function of x in $[a, b[$ for every $u \geq 0$.

Let (Ω, Σ, μ) and the space $L^0(\Omega)$ be defined as in (b). Then the functional

$$\mathcal{I}_\Phi(x, f) = \int_\Omega \Phi(x, |f(t)|)\, d\mu(t)$$

is an Orlicz modular in $L^0(\Omega)$ for every $x \in [a, b[$ (see (b)). We denote by $L_m^0(\Omega)$ the subset of $L^0(\Omega)$ consisting of functions $f \in L^0(\Omega)$ such that $\mathcal{I}_\Phi(\cdot, f\chi_A)$ is Lebesgue measurable in $[a, b[$, for every $A \in \Sigma$, where χ_A is the characteristic function of the set A. In particular, if $\Phi(x, u)$ is a continuous (or a monotone) function of $x \in [a, b[$ for every $u \geq 0$, then $L_m^0(\Omega) = L^0(\Omega)$. We now define an extended functional $\mathcal{A}_\Phi$ on $L_m^0(\Omega)$ by means of the formula

$$\mathcal{A}_\Phi(f) = \sup_{w\in W} \int_a^b a_w(x)\mathcal{I}_\Phi(x, f)\, dm(x), \quad f \in L_m^0(\Omega). \tag{1.5}$$

Then $\mathcal{A}_\Phi$ is a modular on $L_m^0(\Omega)$, and in the case when $\Phi(x, u)$ is a convex function of $u \geq 0$, for all $x \in [a, b[$, $\mathcal{A}_\Phi$ is a convex modular. Supposing there is a notion of convergence in W to an element $w_0 \in W$, strong $\mathcal{A}_\Phi$-summability of f to 0 may be defined by means of the condition

$$\lim_{w\to w_0} \int_a^b a_w(x)\mathcal{I}_\Phi(x, f)\, dm(x) = 0,$$

which gives a connection between the modular $\mathcal{A}_\Phi$ and the notion of strong summability to zero in $L_m^0(\Omega)$.

(f) Let $\mathcal{X}$ be the space of all real-valued functions f on a compact interval $[a, b] \subset \mathbb{R}$ and let φ be a φ-function (see (b)). The *φ-variation* $V_\varphi(f, [a, b])$ of the function f in the interval $[a, b]$ is defined by

$$V_\varphi(f, [a, b]) = \sup_\pi \sum_{j=1}^{n} \varphi(|f(t_j) - f(t_{j-1})|),$$

where the supremum runs over all partitions $\pi = \{a = x_0 < x_1 < \cdots < x_n = b\}$ of the interval $[a, b]$. Then

$$\rho(f) = |f(a)| + V_\varphi(f, [a, b])$$

is a modular on $\mathcal{X}$. The respective modular space $\mathcal{X}_\rho$ is called the *space of functions of bounded φ-variation in* $[a, b]$.

The following statement gives a necessary and sufficient condition for norm convergence of a sequence of functions $f_n \in \mathcal{X}_\rho$ in the sense of the norm $\|\cdot\|_\rho$ (or the F-norm $|||\cdot|||_\rho$):

Theorem 1.2. *Let $\mathcal{X}_\rho$ be the modular space generated by a modular ρ and let $f \in \mathcal{X}_\rho$ and $f_n \in \mathcal{X}_\rho$ for $n = 1, 2, \ldots$ There holds $f_n \to f$ in the sense of the norm $\|\cdot\|_\rho$ (F-norm $|||\cdot|||_\rho$), if and only if $\rho(\lambda(f_n - f)) \to 0$ as $n \to +\infty$, for every $\lambda > 0$.*

Proof (in the case of $\|\cdot\|_\rho$). Let ρ be a convex modular in $\mathcal{X}$ and let $\rho(\lambda(f_n - f)) \to 0$ as $n \to +\infty$ for every $\lambda > 0$. Taking $\lambda = 1/u$ for a fixed $u > 0$ we obtain that there is an index N_u such that $\rho((f_n - f)/u) \leq 1$, for $n > N_u$. This means that $\|f_n - f\|_\rho \leq u$ for $n > N_u$, i.e. $\|f_n - f\|_\rho \to 0$ as $n \to +\infty$. Conversely, let us suppose that $\|f_n - f\|_\rho \to 0$ as $n \to +\infty$. Hence for every $\varepsilon > 0$ and $u > 0$, there exists an index N such that $\|f_n - f\|_\rho < \varepsilon u$, for $n > N$, i.e. $\|(f_n - f)/u\|_\rho < \varepsilon$, for $n > N$. Supposing $\varepsilon \leq 1$, we obtain, by Theorem 1.1 (c),

$$\rho((f_n - f)/u) \leq \|f_n - f\|_\rho < \varepsilon,$$

for $n > N$. Therefore $\rho((f_n - f)/u) \to 0$ as $n \to +\infty$ for every $u > 0$, i.e. $\rho(\lambda(f_n - f)) \to 0$ as $n \to +\infty$, for every $\lambda > 0$.

The proof in the case of $|||\cdot|||_\rho$ is analogous. □

In connection with Theorem 1.2, one may introduce another kind of convergence in a modular space. Namely, we say that a sequence of functions $f_n \in \mathcal{X}_\rho$ is *ρ-convergent*, or *modular convergent* to a function $f \in \mathcal{X}_\rho$, if there exists a $\lambda > 0$ such that $\rho(\lambda(f_n - f)) \to 0$, as $n \to +\infty$; we denote this convergence by $f_n \xrightarrow{\rho} f$ as $n \to +\infty$. Obviously, convergence in the sense of the norm $\|\cdot\|_\rho$ (or F-norm $|||\cdot|||_\rho$) generated by ρ of a sequence (f_n) to f implies its ρ-convergence to f. Both notions are equivalent e.g. in the case given in Example 1.5 (a) or in the case of L^p-spaces. However, they are not equivalent in the general case, as the following example of an Orlicz space shows.

Example 1.6. Let μ be the Lebesgue measure in the σ-algebra Σ_L of Lebesgue measurable subsets of the interval $\Omega = [0, 1]$. Denote by $L^{\varphi}([0, 1])$ the Orlicz space generated by the convex φ-function $\varphi(u) = e^u - u - 1$. Let $A_n =]2^{-n}, 2^{-n+1}]$ for $n = 1, 2, \ldots$ Put $f_n(t) = k/2$, for $t \in A_k$, $k = n, n+1, \ldots$, and $f_n(t) = 0$ for remaining $t \in [0, 1]$, where $n = 1, 2, \ldots$ Obviously, $f_n(t) \to 0$ as $n \to +\infty$ for every $t \in [0, 1[$. Taking $\rho(f) = \int_0^1 \varphi(|f(t)|)\, dt$, we have

$$\rho(f_n) = \left(\frac{\sqrt{e}}{2}\right)^n \left(1 - \frac{\sqrt{e}}{2}\right)^{-1} - \frac{n+3}{2^n} \to 0, \quad \text{as } n \to +\infty,$$

$$\rho(2f_n) = \sum_{k=n}^{\infty} (e/2)^k - \sum_{k=n}^{\infty} k2^{-k} - \sum_{k=n}^{\infty} 2^{-k} = \infty, \quad \text{for every } n.$$

Hence $f_n \xrightarrow{\rho} 0$, but $\|f_n\|_\rho \not\to 0$ as $n \to +\infty$.

There appears the problem, under what assumptions on the modular ρ, norm convergence and modular convergence are equivalent in a modular space $\mathcal{X}_\rho$. In considering this problem we shall limit ourselves to Orlicz spaces $L^{\varphi}(\Omega, \Sigma, \mu)$ (Example 1.5 (b)).

The crucial property needed here is the (Δ_2)-*condition* for the φ-function φ. There are three versions of this condition: for *all* u, for *large* u and for *small* u; we shall denote them as $(\Delta_2)_a$, $(\Delta_2)_l$ and $(\Delta_2)_s$, respectively. We say that φ satisfies $(\Delta_2)_a$, if there exists a constant $M > 0$ such that the inequality

$$\varphi(2u) \le M\varphi(u) \tag{1.6}$$

holds for all $u \ge 0$. The function φ is said to satisfy $(\Delta_2)_l$, (resp. $(\Delta_2)_s$), if there are constants $M > 0$ and $u_0 \ge 0$ such that the inequality (1.6) holds for $u \ge u_0$ (resp. for $0 \le u \le u_0$). It is easily seen that the condition $(\Delta_2)_l$, (resp $(\Delta_2)_s$), is equivalent to the following one $(\Delta_2)'_l$ (resp. $(\Delta_2)'_s$): for every $u_0 > 0$ there exists a constant $M(u_0) > 0$ such that for every $u \ge u_0$ (resp. for every $0 \le u \le u_0$) there holds the inequality

$$\varphi(2u) \le M(u_0)\varphi(u).$$

We shall prove the following

Theorem 1.3. *Let $L^{\varphi}(\Omega, \Sigma, \mu)$ be an Orlicz space (see Example 1.5 (b)). Then each of the following conditions is sufficient in order that norm convergence and modular convergence be equivalent in $L^{\varphi}(\Omega, \Sigma, \mu)$.*

1. *φ satisfies $(\Delta_2)_a$,*
2. *φ satisfies $(\Delta_2)_l$ and $\mu(\Omega) < +\infty$,*
3. *φ satisfies $(\Delta_2)_s$ and $\Omega = \mathbb{N}$, μ being the counting measure in $\mathbb{N}$.*

Proof. Obviously, norm convergence of (f_n) to f is equivalent to the condition $I_\varphi(2^N\lambda(f_n - f)) \to 0$ as $n \to +\infty$, for some $\lambda > 0$ and all $N = 1, 2, \dots$ Suppose that there holds the condition 1. and that $f_n \xrightarrow{I_\varphi} f$. Then there exists a $\lambda > 0$ such that $I_\varphi(\lambda(f_n - f)) \to 0$ as $n \to +\infty$. The condition 1. implies, by easy induction, that $\varphi(2^N u) \le M^N \varphi(u)$, for all $u \ge 0$. Hence

$$I_\varphi(2^N\lambda(f_n - f)) \le M^N I_\varphi(\lambda(f_n - f)) \to 0, \quad \text{as } n \to +\infty.$$

Consequently, $f_n \to f$ in the sense of the norm in $L^\varphi(\Omega, \Sigma, \mu)$.

Now let us suppose 2. to be satisfied and $f_n \xrightarrow{I_\varphi} f$, i.e. $I_\varphi(\lambda(f_n - f)) \to 0$ as $n \to +\infty$ for some $\lambda > 0$. First, let us remark that the condition $(\Delta_2)_l$ implies, by an easy induction, the following one: for every $u_0 > 0$ and every natural number N, there exists an $M_N(u_0) > 0$ such that for every $u \ge u_0$ there holds the inequality

$$\varphi(2^N u) \le M_N(u_0)\varphi(u).$$

Let us denote $A_n(u_0) = \{t \in \Omega : |f_n(t) - f(t)| > u_0\}$, for $u_0 > 0$, $n = 1, 2, \dots$. Then

$$\begin{aligned} I_\varphi(2^N\lambda(f_n - f)) &\le \int_{A_n(u_0)} \varphi(2^N\lambda|f_n(t) - f(t)|)\, d\mu(t) + \varphi(2^N\lambda u_0)\mu(\Omega) \\ &\le M_N(u_0) \int_{A_n(u_0)} \varphi(\lambda|f_n(t) - f(t)|)\, d\mu(t) + \varphi(2^N\lambda u_0)\mu(\Omega) \\ &\le M_N(u_0) I_\varphi(\lambda(f_n - f)) + \varphi(2^N\lambda u_0)\mu(\Omega). \end{aligned}$$

Let us choose an arbitrary $\varepsilon > 0$. We may find $u_0 > 0$ so small that $\varphi(2^N\lambda u_0)\mu(\Omega) < \varepsilon/2$. Since $I_\varphi(\lambda(f_n - f)) \to 0$ as $n \to +\infty$, so keeping u_0 fixed we may find an index n_0 such that $M_N(u_0) I_\varphi(\lambda(f_n - f)) < \varepsilon/2$ for $n \ge n_0$. Thus, $I_\varphi(2^N\lambda(f_n - f)) < \varepsilon$ for $n \ge n_0$. This shows that $f_n \to f$ in the sense of the norm in $L^\varphi(\Omega, \Sigma, \mu)$.

The proof that if $f_n \xrightarrow{I_\varphi} f$ as $n \to +\infty$ and there holds 3., then $f_n \to f$ in the sense of the norm in ℓ^φ, is obtained in a similar manner applying the fact that every sequence $x \in \ell^\varphi$ is bounded. We leave the details to the reader. □

1.3 Quasiconvex modulars

We are going to extend the notion of convexity of a modular ρ and of a φ-function φ to the more general case of quasiconvexity. Let ρ be a modular on a real linear function space $\mathcal{X}$. We say that ρ is *quasiconvex with a constant* $M \ge 1$, if for any natural number n and for every elements $f_1, f_2, \dots, f_n \in \mathcal{X}$ and nonnegative numbers $\alpha_1, \alpha_2, \dots, \alpha_n$, satisfying the condition $\alpha_1 + \dots + \alpha_n = 1$ there holds the inequality

$$\rho\Big(\sum_{j=1}^{n} \alpha_j f_j\Big) \le M \sum_{j=1}^{n} \alpha_j \rho(M f_j).$$

Let us remark that if $M = 1$, this is equivalent to convexity of the modular ρ.

Example 1.7. (a) In case when $\mathcal{X} = \mathbb{R}$, i.e. $\mathcal{X}$ is the space of all constant functions on any set Ω, we have $\rho : \mathbb{R} \to \widetilde{\mathbb{R}}_0^+$, and we may write φ in place of ρ. This yields the following definition: a function $\varphi : \mathbb{R} \to \widetilde{\mathbb{R}}_0^+$ is called *quasiconvex with a constant* $M \geq 1$ (and in the case of $M = 1$, convex), if the conditions $u_1, u_2, \ldots, u_n \in \mathbb{R}$, $\alpha_1, \alpha_2, \ldots, \alpha_n \geq 0$ and $\alpha_1 + \alpha_2 + \cdots + \alpha_n = 1$ imply the inequality

$$\varphi\Big(\sum_{j=1}^{n} \alpha_j u_j\Big) \leq M \sum_{j=1}^{n} \alpha_j \varphi(M u_j)$$

with some fixed $M \geq 1$, independent of n.

(b) Let $\varphi : \mathbb{R} \to \mathbb{R}_0^+$ be any even function such that $u^2/2 \leq \varphi(u) \leq u^2$ for all $u \in \mathbb{R}$. Let $u_1, u_2, \ldots, u_n \geq 0$, $\alpha_1, \alpha_2, \ldots, \alpha_n \geq 0$ and $\alpha_1 + \alpha_2 \cdots + \alpha_n = 1$. Then there holds, by convexity of the function $\varphi(u) = u^2$,

$$\begin{aligned}\varphi\Big(\sum_{j=1}^{n} \alpha_j u_j\Big) &\leq \Big(\sum_{j=1}^{n} \alpha_j u_j\Big)^2 \leq \sum_{j=1}^{n} \alpha_j u_j^2 \\ &= \sqrt[3]{2} \sum_{j=1}^{n} \alpha_j \frac{1}{2} (\sqrt[3]{2} u_j)^2 \leq \sqrt[3]{2} \sum_{j=1}^{n} \alpha_j \varphi(\sqrt[3]{2} u_j).\end{aligned}$$

Hence φ is quasiconvex with the constant $M = \sqrt[3]{2}$. Obviously φ does not need to be convex.

Remark 1.1. In the definition of quasiconvexity, as well as in the definition of convexity, one may replace the equality $\alpha_1 + \alpha_2 + \cdots + \alpha_n = 1$ by the inequality $\alpha_1 + \alpha_2 + \cdots + \alpha_n \leq 1$. Indeed, let us suppose ρ to be quasiconvex with a constant $M \geq 1$ and let $\beta_1, \beta_2, \ldots, \beta_n \geq 0$, $\beta_1 + \beta_2 + \cdots + \beta_n < 1$. Let us put $\alpha_j = \beta_j$ for $j = 1, 2, \ldots, n$ and $\alpha_{n+1} = 1 - (\beta_1 + \beta_2 + \cdots + \beta_n)$, then $\alpha_1 + \alpha_2 + \cdots + \alpha_n + \alpha_{n+1} = 1$. Hence for any $f_1, f_2, \ldots, f_n \in \mathcal{X}$ we have

$$\rho\Big(\sum_{j=1}^{n+1} \alpha_j f_j\Big) \leq M \sum_{j=1}^{n+1} \alpha_j \rho(M f_j),$$

where $f_{n+1} = 0$. Consequently,

$$\begin{aligned}\rho\Big(\sum_{j=1}^{n} \beta_j f_j\Big) &= \rho\Big(\sum_{j=1}^{n+1} \alpha_j f_j\Big) \\ &\leq M \sum_{j=1}^{n+1} \alpha_j \rho(M f_j) = M \sum_{j=1}^{n} \beta_j \rho(M f_j).\end{aligned}$$

Let ρ be a modular on a linear subspace $\mathcal{X}$ of the space $L^0(\Omega)$. The modular ρ is called *J-quasiconvex* (*quasiconvex in Jensen's sense*), *with a constant* $M \geq 1$, if for all Σ-measurable functions $p : \Omega \to \mathbb{R}_0^+$ such that $\|p\|_{L^1(\Omega)} = \int_\Omega p(t)\,d\mu(t) = 1$ and for all functions $F : \Omega \times \Omega \to \mathbb{R}_0^+$ such that $F(\cdot, :) \in L^0(\Omega \times \Omega)$ and $F(t, :) \in \mathcal{X}$ for every $t \in \Omega$, there holds the following inequality:

$$\rho\left(\int_\Omega p(t)F(t, :)\,d\mu(t)\right) \leq M \int_\Omega p(t)\rho(MF(t, :))\,d\mu(t),$$

and both sides of this inequality make sense. If $M = 1$, we call ρ *J-convex* (*convex in Jensen's sense*). Similarly as in the discrete case, we may replace the equality $\|p\|_{L^1(\Omega)} = 1$, by the inequality $\|p\|_{L^1(\Omega)} \leq 1$.

Example 1.8. Taking $\mathcal{X} = \mathbb{R}$, we obtain a function φ in place of ρ (see Example 1.7 (a)). This function φ is called *J-quasiconvex* (*quasiconvex in Jensen's sense*), *with a constant* $M \geq 1$, if there holds the inequality

$$\varphi\left(\int_\Omega p(t)F(t)\,d\mu(t)\right) \leq M \int_\Omega p(t)\varphi(MF(t))\,d\mu(t) \tag{1.7}$$

for all Σ-measurable function $p : \Omega \to \mathbb{R}_0^+$ such that $\|p\|_{L^1(\Omega)} = 1$ and all Σ-measurable functions $F : \Omega \to \mathbb{R}_0^+$. In case when $M = 1$ the above inequality becomes the well-known Jensen's inequality for convex functions φ.

Theorem 1.4. *Let (Ω, Σ, μ) be a measure space with a nonatomic measure such that $\mu(\Omega) > 0$ and let $\mathcal{X}$ be a linear subspace of $L^0(\Omega)$ containing characteristic functions of the sets $A \in \Sigma$ of finite measure μ and such that if $f \in \mathcal{X}$ then $|f| \in \mathcal{X}$. Let ρ be a monotone, J-quasiconvex modular on $\mathcal{X}$. Then ρ is quasiconvex in $\mathcal{X}$.*

Proof. First, let us suppose that $\mu(\Omega) \geq 1$. Since μ is nonatomic, one may select a set $C \in \Sigma$ of measure $\mu(C) = 1$. Taking $p(t) = \chi_C(t)$, the characteristic function of the set C, we obtain, by quasiconvexity of ρ, the inequality

$$\rho\left(\int_C F(t, :)d\mu(t)\right) \leq M \int_C \rho(MF(t, :))\,d\mu(t) \tag{1.8}$$

for every nonnegative function $F(\cdot, :) \in L^0(\Omega \times \Omega)$ such that $F(t, :) \in \mathcal{X}$ for all $t \in \Omega$. Let $\alpha_1, \alpha_2, \ldots, \alpha_n \geq 0$, $\alpha_1 + \alpha_2 + \cdots + \alpha_n = 1$. Since μ is atomless, there exist pairwise disjoint subsets $A_1, A_2, \ldots, A_n \in \Sigma$ of the set C such that $A_1 \cup A_2 \cup \cdots \cup A_n = C$ and $\mu(A_j) = \alpha_j$ for $j = 1, 2, \ldots, n$. Let $f_1, f_2, \ldots, f_n \in \mathcal{X}$ and let

$$F(t, s) = \chi_{A_1}(t)f_1(s) + \cdots + \chi_{A_n}(t)f_n(s)$$

for $s, t \in \Omega$. Obviously, $F(\cdot, :) \in L^0(\Omega \times \Omega)$ and since $\mathcal{X}$ is a linear space, we have $F(t, :) \in \mathcal{X}$ for all $t \in \Omega$. Since $f \in \mathcal{X}$ implies $|f| \in \mathcal{X}$, there also holds

$|F(t,:)| \in \mathcal{X}$ for $t \in \Omega$. Hence, applying monotonicity of ρ and the inequality (1.8), we obtain

$$\begin{aligned}\rho\Big(\sum_{j=1}^{n}\alpha_j f_j\Big) &= \rho\Big(\sum_{j=1}^{n}\mu(A_j) f_j\Big) = \rho\left(\int_C F(t,:)\,d\mu(t)\right) \\ &\le \rho\left(\int_C |F(t,:)|\,d\mu(t)\right) \le M\int_C \rho(M|F(t,:)|)\,d\mu(t) \\ &= M\int_C \sum_{j=1}^{n}\rho(M|f_j(:)|)\chi_{A_j}(t)\,d\mu(t) = M\sum_{j=1}^{n}\alpha_j\rho(M|f_j|) \\ &= M\sum_{j=1}^{n}\alpha_j\rho(Mf_j).\end{aligned}$$

Thus, ρ is quasiconvex with the constant M.

Now, suppose that $0 < \mu(\Omega) < 1$. Put $\nu(A) = \mu(A)/\mu(\Omega)$ for $A \in \Sigma$, then $\nu(\Omega) = 1$. Let $F : \Omega \times \Omega \to \mathbb{R}_0^+$ be such that $F(\cdot,:) \in L^0(\Omega \times \Omega)$ and $F(t,:) \in \mathcal{X}$ for every $t \in \Omega$ and let $p : \Omega \to \mathbb{R}_0^+$, $\|p\|_{L^1(\Omega)} = 1$. Then we have

$$\begin{aligned}\rho\left(\int_\Omega p(t)F(t,:)\,d\nu(t)\right) &= \rho\left(\int_\Omega p(t)\frac{F(t,:)}{\mu(\Omega)}\,d\mu(t)\right) \\ &\le M\int_\Omega p(t)\rho\left(\frac{M}{\mu(\Omega)}F(t,:)\right)d\mu(t) \\ &\le M'\int_\Omega p(t)\rho(M'F(t,:))\,d\nu(t),\end{aligned}$$

where $M' = M/\mu(\Omega)$. Thus, by the first part of the proof, ρ is quasiconvex with the constant $M' = M/\mu(\Omega)$. □

The converse problem, under what condition quasiconvexity of ρ implies its J-quasiconvexity, will be examined in the case of the modular generating an Orlicz space (see Example 1.5 (b)), i.e., $\rho(f)$ is defined by the formula (1.3).

Example 1.9. (a) Let $\varphi : \mathbb{R} \to \mathbb{R}_0^+$ be a nondecreasing function in $\mathbb{R}^+$ such that φ is quasiconvex with a constant $M \ge 1$. Then it is J-quasiconvex (see Example (1.8)), i.e. there holds the inequality (1.7). We prove it first in the case when both p and F are simple functions. Then there are constants $a_1, a_2, \dots, a_n \ge 0$, $b_1, b_2, \dots b_n \ge 0$ and pairwise disjoint sets $A_1, A_2, \dots, A_n \in \Sigma$ with $A_1 \cup A_2 \cup \dots \cup A_n = \Omega$ such that $p(t) = \sum_{i=1}^n a_i\chi_{A_i}(t)$ and $F(t) = \sum_{j=1}^n b_j\chi_{A_j}(t)$, where $a_j = 0$ if $\mu(A_j) = +\infty$ and we put then $a_j\mu(A_j) = 0$, by convention. Since $\|p\|_{L^1(\Omega)} = 1$, we have

$\sum_{j=1}^{n} a_j \mu(A_j) = 1$. Hence we obtain, by quasiconvexity of φ with a constant $M \geq 1$,

$$\begin{aligned}\varphi\Big(\int_\Omega p(t)F(t)\,d\mu(t)\Big) &= \varphi\Big(\sum_{j=1}^{n} a_j\mu(A_j)b_j\Big)\\ &\leq M\sum_{j=1}^{n} a_j\mu(A_j)\varphi(Mb_j)\\ &= M\int_\Omega p(t)\varphi(MF(t))\,d\mu(t),\end{aligned}$$

i.e. the inequality (1.7). By the remark after the definition of J-quasiconvexity with $M \geq 1$, the same holds if $\|p\|_{L^1(\Omega)} \leq 1$. Now, let p and F be arbitrary nonnegative, Σ-measurable functions with $\|p\|_{L^1(\Omega)} = 1$. Then there are two sequences (p_n) and (F_n) of nonnegative simple functions such that $p_n(t) \nearrow p(t)$ and $F_n(t) \nearrow F(t)$ for every $t \in \Omega$. Obviously, $\|p_n\|_{L^1(\Omega)} \leq 1$. Applying the inequality (1.7) to the functions p_n and F_n, we get

$$\begin{aligned}\varphi\left(\int_\Omega p_n(t)F_n(t)\,d\mu(t)\right) &\leq M\int_\Omega p_n(t)\varphi(MF_n(t))\,d\mu(t)\\ &\leq M\int_\Omega p(t)\varphi(MF(t))\,d\mu(t).\end{aligned}$$

Passing to the limit at the left-hand side of the above inequality and applying the Beppo Levi's theorem, we easily obtain the inequality (1.7) for arbitrary p and F. Thus φ is J-quasiconvex with the same constant M.

(b) Let ρ be the modular (1.3) generating the Orlicz space (Example 1.5 (b)), where we suppose that Σ contains a set C of finite, positive measure. Let us suppose ρ to be quasiconvex with a constant $M \geq 1$. We show that the function φ, generating the modular ρ, is also quasiconvex with the same constant. Indeed, let $\alpha_1, \alpha_2, \ldots \alpha_n \geq 0$, $\alpha_1 + \alpha_2 + \cdots + \alpha_n = 1$, and $u_1, u_2, \ldots, u_n \in \mathbb{R}$. Let $f_j(t) = u_j\chi_C(t)$, then $f_j \in L^0(\Omega)$, $j = 1, 2, \ldots, n$. By quasiconvexity of ρ, we have

$$\begin{aligned}\mu(C)\varphi\Big(\sum_{j=1}^{n}\alpha_j u_j\Big) &= \int_\Omega \varphi\Big(\sum_{j=1}^{n}\alpha_j f_j(t)\Big)\,d\mu(t) = \rho\Big(\sum_{j=1}^{n}\alpha_j f_j\Big)\\ &\leq M\sum_{j=1}^{n}\alpha_j\rho(Mf_j) = M\sum_{j=1}^{n}\alpha_j\int_\Omega \varphi(Mf_j(t))\,d\mu(t)\\ &= M\mu(C)\sum_{j=1}^{n}\alpha_j\varphi(Mu_j)\end{aligned}$$

which proves φ to be quasiconvex with constant M.

Now by Example 1.9 (a) we conclude that φ is J-quasiconvex with constant M. But this implies that the modular ρ is also quasiconvex with constant M, because, by Fubini's theorem, we have

$$\begin{aligned}\rho\left(\int_\Omega p(t)F(t,:)\,d\mu(t)\right) &= \int_\Omega \varphi\left(\int_\Omega p(t)F(t,s)\,d\mu(t)\right)d\mu(s)\\ &\le \int_\Omega M\left(\int_\Omega p(t)\varphi(MF(t,s))\,d\mu(t)\right)d\mu(s)\\ &= M\int_\Omega p(t)\left(\int_\Omega \varphi(MF(t,s))\,d\mu(s)\right)d\mu(t)\\ &= M\int_\Omega p(t)\rho(MF(t,:))\,d\mu(t).\end{aligned}$$

1.4 Subbounded modulars

For further considerations we need the notion of subboundedness of a modular η in a linear subspace $\mathcal{X}$ of $L^0(\Omega)$, where Ω is provided with an operation $+ : \Omega\times\Omega\to\Omega$. We will assume that $\mathcal{X}$ is *invariant* with respect to the operation +, i.e. if $f\in\mathcal{X}$ then $f(t+\cdot)\in\mathcal{X}$ for every $t\in\Omega$. Such a modular η is called *subbounded* (with respect to the operation +), if there exist a constant $C\ge 1$ and a function $\ell:\Omega\to\mathbb{R}_0^+$, $\ell\in L^\infty(\Omega)$, such that for every function $f\in\mathcal{X}$ and for every $t\in\Omega$ there holds the inequality

$$\eta[f(t+\cdot)]\le\eta(Cf)+\ell(t),$$

and its left-hand side is Σ-measurable. If the last inequality holds with the function $\ell(t)=0$, for $t\in\Omega$ and $f\in\mathcal{X}$, η is called *strongly subbounded.*

Example 1.10. (a) Let $\mathcal{X}$ be an invariant subspace of $L^0(\Omega)$. If η is *invariant* with respect to the operation +, i.e. $\eta[f(t+\cdot)]=\eta(f)$, for all $f\in\mathcal{X}$ and $t\in\Omega$, then obviously, η is strongly subbounded. This holds in the case of an Orlicz space $L^\varphi(\Omega)$ (Example 1.5 (b)), if $(\Omega,+)$ is a unimodular, locally compact Hausdorff topological group, with Haar measure μ.

(b) Let $(\Omega,+)$ be the same as in (a) and let ρ be the modular η as in Example 1.5 (b), i.e.

$$\eta(f)=\int_\Omega\varphi(t,|f(t)|)\,d\mu(t).$$

Let us suppose that φ satisfies the inequality

$$\varphi(s-t,u)\le\varphi(s,Cu)+h(s,t), \tag{1.9}$$

for all $s, t \in \Omega, u \geq 0$, with some constant $C \geq 1$, independent of s, t, u and $0 \leq h(\cdot, t) \in L^1(\Omega)$, for $t \in \Omega$, $\operatorname{ess\,sup}_t \|h(\cdot, t)\|_{L^1(\Omega)} < +\infty$. Then

$$\begin{aligned} \eta[f(t+\cdot)] &= \int_\Omega \varphi(s, |f(t+s)|)\, d\mu(s) \\ &= \int_\Omega \varphi(s-t, |f(t)|)\, d\mu(s) \\ &\leq \int_\Omega \varphi(s, C|f(s)|)\, d\mu(s) + \int_\Omega h(s,t)\, d\mu(s) \\ &= \eta(Cf) + \ell(t), \end{aligned}$$

with $\ell(\cdot) = \int_\Omega h(s, \cdot)\, d\mu(s) \in L^\infty(\Omega)$. Thus η is subbounded.

In many problems we shall need a connection between two modulars ρ, η on $L^0(\Omega)$, and a function $\psi : \Omega \times \mathbb{R}_0^+ \to \mathbb{R}_0^+$ satisfying the following conditions: $\psi(\cdot, u)$ is Σ-measurable for all $u \geq 0$, $\psi(t, :)$ is continuous and nondecreasing for every $t \in \Omega$, $\psi(t, 0) = 0$, $\psi(t, u) > 0$ for $u > 0$, $\psi(t, u) \to +\infty$ as $u \to +\infty$, for all $t \in \Omega$.

We say that $\{\rho, \psi, \eta\}$ is a *properly directed triple* if there is a set $\Omega_0 \subset \Omega$, $\Omega_0 \in \Sigma$ such that $\mu(\Omega \setminus \Omega_0) = 0$ and for every λ with $0 < \lambda < 1$ there exists a C_λ, $0 < C_\lambda < 1$, satisfying the inequality

$$\rho[C_\lambda \psi(t, |F(\cdot)|)] \leq \eta(\lambda F(\cdot)),$$

for all $t \in \Omega_0$ and $F \in L^0(\Omega)$. Let us remark that one may choose C_λ in such a manner that $C_\lambda \searrow 0$ as $\lambda \searrow 0$. Moreover, the above condition immediately implies the following inequality:

$$\rho[C_\lambda \psi(t, |F_t(\cdot)|)] \leq \eta(\lambda F_t(\cdot))$$

for every $t \in \Omega_0$ and for any family $(F_t(\cdot))_{t \in \Omega_0}$ of functions $F_t \in L^0(\Omega)$.

Example 1.11. Let ρ be an Orlicz modular on $L^0(\Omega)$, i.e.

$$\rho(f) = \int_\Omega \varphi(|f(t)|)\, d\mu(t),$$

where φ is a convex φ-function. Let ψ be the inverse to φ and $0 < \lambda < 1$. Then ψ is concave and there holds

$$\begin{aligned} \rho[\lambda\psi(|F(\cdot)|)] &= \int_\Omega \varphi[\lambda\psi(|F(s)|)]\, d\mu(s) \\ &\leq \int_\Omega \varphi[\psi(\lambda|F(s)|)]\, d\mu(s) \\ &= \int_\Omega \lambda|F(s)|\, d\mu(s) = \eta(\lambda F), \end{aligned}$$

if we take $\eta(F) = \|F\|_{L^1(\Omega)}$. This means that $\{\rho, \psi, \eta\}$ is a properly directed triple with $C_\lambda = \lambda$.

1.5 Bibliographical notes

As regards linear kernel functions and their applications in approximation theory, see the fundamental monograph by P. L. Butzer and R. J. Nessel, [67]. The ideas originating the method used in our book may be found in J. Musielak [150], in the case of linear kernel functions and in J. Musielak [152] in the case of convolution-type nonlinear kernel functions, with Lebesgue measure on an interval in $\mathbb{R}$. A study of Urysohn and Hammerstein operators may be found in M.A. Krasnoselskii and Y. B. Rutickii [133]. See also [132], [131], [113], [182], [193], [194], [195].

The notion of a modular was introduced by H. Nakano [167] in 1951. The definition of a modular ρ (not necessarily convex) is due to J. Musielak and W. Orlicz [164], 1959, together with Theorem 1.1. The Orlicz space (Example 1.5 (b)) was introduced by W. Orlicz [170], [171] in 1932 and 1936. For the theory of Orlicz spaces we refer to [133], [140], [174]. The generalized Orlicz space or Musielak–Orlicz space was introduced by H. Nakano [166] and developed in [164]. As regards the notion of a modular, modular space and Examples 1.1 (a)–(d), detailed comments may be found in [153] or in [131] and for Example 1.1 (e) see [18]. In [131], W. M. Kozlowski defined a *function modular* $\rho : \Sigma \times \Sigma \to [0, +\infty]$, where Σ is the class of simple functions on Ω with support of finite measure; ρ defines a *modular function space* in the sense of [131]. These ideas were developed in [126], [125], and recently in [92], [93] and [94]. The notion of modular convergence and Theorem 1.2 were given in [164], and condition Δ_2 together with Theorem 1.3 were obtained by W. Orlicz [170], [171].

Quasiconvexity of a function φ was investigated by A. Gogatishvili and V. Kokilashvili [112] (1993), and in the book of V. Kokilashvili and M. Krbec [129] (1991), and further applied by I. Mantellini and G. Vinti [142], to estimates of nonlinear integral operators. Quasiconvex modulars ρ were defined in [21] (1998) and applied in order to obtain some modular inequalities related to the Fubini–Tonelli theorem.

Inequality (1.9) for a function φ with parameter was introduced by A. Kamińska [123] and by A.Kamińska and R. Płuciennik [124] as a necessary and sufficient condition in order that $\|f(\cdot + h) - f(\cdot)\|_{I_\varphi} \to 0$ as $h \to 0$ for $f \in L^\varphi(\Omega)$, $\Omega \subset \mathbb{R}^n$. Inequality (1.9) was used in [19] in order to define, in the case of a group $(\Omega, +)$, a stronger property than subboundedness of a modular, requiring that $h(t) \to 0$ as $t \to \theta$ (we call this stronger property "boundedness" of the modular: see the text after the proof of Theorem 2.3). Strongly subbounded modulars were defined in [160], in connection with problems of best approximation of periodic functions by means of trigonometric polynomials in modular spaces. The notion of a properly directed triple was introduced in [34] in the case of a group $(\Omega, +)$ and further developed in [19], [22].

Chapter 2

Absolutely continuous modulars and moduli of continuity

2.1 Absolutely finite and absolutely continuous modulars

In this chapter we start with formulating a modular version of the Lebesgue dominated convergence theorem. This requires further properties of modulars. Let us recall that a modular ρ on $\mathcal{X} \subset L^0(\Omega)$ is called *monotone*, if $f, g \in \mathcal{X}$ and $|f| \leq |g|$ imply $\rho(f) \leq \rho(g)$ (see Section 1.2). We say that a modular ρ is *finite*, if $\chi_A \in \mathcal{X}_\rho$ for every $A \in \Sigma$ such that $\mu(A) < +\infty$. A modular ρ on $\mathcal{X}$ is said to be *absolutely finite*, if it is finite and if for every $\varepsilon > 0$ and every $\lambda_0 > 0$ there exists a $\delta > 0$ such that every set $B \in \Sigma$ with $\mu(B) < \delta$ satisfies the inequality $\rho(\lambda_0 \chi_B) < \varepsilon$.

Example 2.1. (a) Let ρ be the modular I_φ generating the Orlicz space $L^0_\rho(\Omega) = L^\varphi(\Omega)$ (Example 1.5 (b)). Obviously, ρ is monotone and absolutely finite in $L^\varphi(\Omega)$.

(b) Let ρ be the modular I_φ generating a generalized Orlicz space $L^0_\rho(\Omega) = L^\varphi(\Omega)$ (Example 1.5 (c)). It is easily observed that ρ is always monotone, ρ is finite if and only if $\varphi(\cdot, u)$ is *locally integrable for small u* (i.e. for every $A \in \Sigma$ with $\mu(A) < +\infty$ there is a $u > 0$ such that $\int_A \varphi(t, u)\, d\mu(t) < +\infty$), and ρ is absolutely finite if and only if $\varphi(\cdot, u)$ is *locally integrable* (i.e. for every $A \in \Sigma$ with $\mu(A) < +\infty$ there holds $\int_A \varphi(t, u)\, d\mu(t)$ for all $u > 0$).

(c) Let $\mathcal{X}$ be the space of all real-valued functions on the interval $[0, 1] \subset \mathbb{R}$ and let $V_0^1(f)$ be the classical (Jordan) variation of a function $f \in \mathcal{X}$ on $[0, 1]$. Let $\rho(f) = |f(0)| + V_0^1(f)$ (Example (1.5)(f), with $\varphi(u) = |u|$). Take $f(t) = \sin(\pi t)$ and $g(t) = 1$, for $t \in [0, 1]$. Then $0 \leq f(t) \leq g(t)$ for $t \in [0, 1]$, but $\rho(f) = 2$ and $\rho(g) = 1$, whence $\rho(f) > \rho(g)$. Thus ρ is not monotone.

Let $A_1, A_2, \ldots,$ be a sequence of pairwise disjoint, closed subintervals of the interval $[0, 1]$ and let $A = A_1 \cup A_2 \cup \cdots$, and μ- the Lebesgue measure on $[0, 1]$. Then $\mu(A) < +\infty$, but $\chi_A \notin \mathcal{X}_\rho$, because $V_0^1(\chi_A) = +\infty$. Thus ρ is not finite. However, if we take as μ the counting measure, then $\mu(A) < +\infty$ if and only if $A \in [0, 1]$ is a finite set and so the modular ρ becomes finite.

In the following we shall need some additional assumptions on the linear subspace $\mathcal{X} \subset L^0(\Omega)$. We say that $\mathcal{X}$ is a *correct subspace* of $L^0(\Omega)$, if

(a) $A \in \Sigma$ and $\mu(A) < +\infty$ imply $\chi_A \in \mathcal{X}$,

(b) $f \in \mathcal{X}$ and $A \in \Sigma$ imply $f\chi_A \in \mathcal{X}$.

Obviously, $\mathcal{X} = L^0(\Omega)$ is a correct subspace of itself. Moreover, if $\mathcal{X}$ is a correct subspace of $L^0(\Omega)$ and $f \in \mathcal{X}$, then $|f| \in \mathcal{X}$. Indeed, let $f \in \mathcal{X}$ and let $A = \{t \in \Omega : f(t) \geq 0\}$, $B = \Omega \setminus A$. Then $A, B \in \Sigma$, and so $f_+ = f\chi_A \in \mathcal{X}$ and $f_- = f\chi_B \in \mathcal{X}$. Hence $|f| = f_+ + f_- \in \mathcal{X}$.

Let $\mathcal{X}$ be a linear correct subspace of $L^0(\Omega)$. We say that ρ is *absolutely continuous* (a.c.), if there exists an $\alpha > 0$ such that every function $f \in \mathcal{X}$ with $\rho(f) < +\infty$ satisfies the following two conditions:

(a) for every $\varepsilon > 0$ there exists a set $A \in \Sigma$ with $\mu(A) < +\infty$ such that $\rho(\alpha f\chi_{\Omega\setminus A}) < \varepsilon$,

(b) for every $\varepsilon > 0$ there exists $\delta > 0$ such that, for every set $B \in \Sigma$ with $\mu(B) < \delta$, there holds $\rho(\alpha f\chi_B) < \varepsilon$.

If $\mu(\Omega) < +\infty$ then the condition (a) is obviously satisfied.

Example 2.2. (a) Let ρ be the modular I_φ in $L^0(\Omega)$ generating the generalized Orlicz space (Example 1.5 (c)). The condition $\rho(f) < +\infty$ means that the function $F(t) = \varphi(t, |f(t)|)$ is integrable on Ω. Then the absolute continuity of the modular ρ with $\alpha = 1$ follows from the well-known properties of the integral.

(b) This example is an exercise in technical problems concerning absolute continuity and may be omitted.

We shall examine the modular $\mathcal{A}_\Phi$ (Example 1.5 (e), formula (1.5)), keeping the notations of Example 1.5 (e). We suppose additionally, that $\int_a^b a_w(x)\,dm(x) \leq 1$, for $w \in W$ and that if $0 \leq g(x) \nearrow s \in \widetilde{\mathbb{R}^+}$ as $x \to b^-$, being $g : [a, b] \to \mathbb{R}_0^+$ a nondecreasing function, then $\int_a^b a_w(x)g(x)\,dm(x) \to s$ for $w \to w_0$. As regards $\Phi : [a, b[\times\mathbb{R}_0^+ \to \mathbb{R}_0^+$ satisfying the conditions 1)–4) of Example 1.5 (e), we also need some additional assumptions. We suppose that there is a $c \in [a, b[$ such that Φ is of *monotone type* in $[c, b[$, i.e. there exist two disjoint sets $R_1, R_2 \subset \mathbb{R}_0^+$ with $R_1 \cup R_2 = \mathbb{R}_0^+$, such that

1°. $\Phi(x, u)$ is a nonincreasing function of $x \in [c, b[$, for every $u \in R_1$,

2°. $\Phi(x, u)$ is a nondecreasing function of $x \in [c, b[$, for every $u \in R_2$.

Finally, we suppose the following *condition* (H) to be satisfied:

(H) For every $f \in L_m^0(\Omega)$ such that $\mathcal{A}_\Phi(f) < +\infty$ and for every $x \in [a, c]$ there is a neighbourhood U_x of x in $[a, c[$ such that the function

$$H_x(\cdot) = \sup_{y \in U_x} \Phi(y, |f(\cdot)|)$$

is μ-integrable over Ω for $x \in [a, c]$.

Let us mention that taking in the condition (H), $U_x = [a, c]$ for all $x \in [a, c]$ one obtains a condition, equivalent to (H). Condition (H) is satisfied for example if we suppose that there are constants $M > 0$ and $u_0 > 0$ such that for all $y, z \in [a, c]$ and $u \geq u_0$, there holds the inequality $\Phi(y, u) \leq M\Phi(z, u)$. The reader is encouraged to go through the details.

Under the above conditions, $\mathcal{A}_\Phi$ is absolutely continuous (a.c.) with respect to the measure μ. Indeed, let $f \in L^0_m(\Omega)$ be such that $\mathcal{A}_\Phi < +\infty$ and let us take an arbitrary set $P \in \Sigma$. Obviously, we have

$$\mathcal{A}_\Phi(f\chi_P) \leq \Gamma_1(f\chi_P) + \Gamma_2(f\chi_P),$$

where

$$\Gamma_1(f) = \sup_{w\in W} \int_a^c a_w(x)\mathcal{I}(x, f)\, dm(x), \quad \Gamma_2(f) = \sup_{w\in W} \int_c^b a_w(x)\mathcal{I}(x, f)\, dm(x).$$

In order to prove $\mathcal{A}_\Phi$ to be a.c. it is sufficient to show both Γ_1 and Γ_2 to be a.c.. First, we prove Γ_1 to be a.c. Applying the condition (H) we observe that the function $H(t) = \sup_{y\in[a,c]} \Phi(y, |f(t)|)$ for $t \in \Omega$ is μ-integrable on Ω. Hence there are a set $S_1 \in \Sigma$ such that $\mu(S_1) < +\infty$ and a number $\delta_1 > 0$ such that

$$\int_{\Omega\setminus S_1} H(t)\, d\mu(t) < \varepsilon/2, \qquad \int_S H(t)\, d\mu(t) < \varepsilon/2, \quad \text{if } S \in \Sigma,\ \mu(S) < \delta_1.$$

Consequently, we have for arbitrary $y \in [a, c]$

$$\int_{\Omega\setminus S_1} \Phi(y, |f(t)|)\, d\mu(t) < \varepsilon/2,$$
$$\int_S \Phi(y, |f(t)|)\, d\mu(t) < \varepsilon/2, \quad \text{if } S \in \Sigma,\ \mu(S) < \delta_1.$$

This implies

$$\mathcal{I}_\Phi(y, f\chi_{\Omega\setminus S_1}) = \int_{\Omega\setminus S_1} \Phi(y, |f(t)|)\, d\mu(t) < \varepsilon/2,$$
$$\mathcal{I}_\Phi(y, f\chi_S) = \int_S \Phi(y, |f(t)|)\, d\mu(t) < \varepsilon/2, \quad \text{if } S \in \Sigma,\ \mu(S) < \delta_1.$$

Thus, we have for any $w \in W$,

$$\int_a^c a_w(x)\mathcal{I}_\Phi(x, f\chi_{\Omega\setminus S_1})\, dm(x) < \frac{\varepsilon}{2} \int_a^c a_w(x)\, dm(x) \leq \varepsilon/2,$$

where $\mu(S_1) < +\infty$, and similarly

$$\int_a^c a_w(x)\mathcal{I}_\Phi(x, f\chi_S)\, dm(x) < \varepsilon/2,$$

if $S \in \Sigma$ and $\mu(S) < \delta_1$. Hence it follows that

$$\Gamma_1(f\chi_{\Omega\setminus S_1}) < \varepsilon/2, \Gamma_1(f\chi_S) < \varepsilon/2, \quad \text{if } S \in \Sigma,\ \mu(S) < \delta_1,$$

where $S_1 \in \Sigma$, $\mu(S_1) < +\infty$. Thus, Γ_1 is a.c..

In order to prove that Γ_2 is a.c., let us write for any $f \in L^0_m(\Omega)$ such that $\mathcal{A}_\Phi(f) < +\infty$, $A = \{t \in \Omega : |f(t)| \in R_1\}$, $B = \{t \in R_2 : |f(t)| \in R_2\}$, where R_1 and R_2 are sets from the definition of the function Φ of monotone type. Let $P \in \Sigma$ be fixed and let $t \in A$. We have $\Phi(x, |f(t)|) \searrow$ as $c \le x \nearrow b^-$, and so $\Phi(x, |f(t)|) \le \Phi(c, |f(t)|)$ for $c \le x < b, t \in A$. Hence

$$\mathcal{I}_\Phi(x, f\chi_A\chi_P) \le \mathcal{I}_\Phi(c, f\chi_A\chi_P) \le \mathcal{I}_\Phi(c, f) < +\infty,$$

for $c \le x < b$; the condition $\mathcal{I}_\Phi < +\infty$ follows, applying the condition (H) to $y = c$. Therefore there exist a set $S_2 \in \Sigma$ with $\mu(S_2) < +\infty$ and a number $\delta_2 > 0$ such that

$$\mathcal{I}_\Phi(x, f\chi_{\Omega\setminus S_2}\chi_A) \le \mathcal{I}_\Phi(c, f\chi_{\Omega\setminus S_2}\chi_A) < \varepsilon/4$$

and

$$\mathcal{I}_\Phi(x, f\chi_S\chi_A) \le \mathcal{I}_\Phi(c, f\chi_S\chi_A) < \varepsilon/4 \quad \text{if } S \in \Sigma,\ \mu(S) < \delta_2,$$

for every $x \in [c, b[$. Thus

$$\begin{aligned} &\Gamma_2(f\chi_{\Omega\setminus S_2}\chi_A) < \varepsilon/4 \\ &\Gamma_2(f\chi_S\chi_A) < \varepsilon/4 \quad \text{if } S \in \Sigma,\ \mu(S) < \delta_2. \end{aligned} \tag{2.1}$$

Next, we have for every $P \in \Sigma$, (see Example 1.5 (e))

$$\Phi(x, |f(t)|\chi_P(t)\chi_B(t)) \nearrow \widetilde{\Phi}(|f(t)|\chi_P(t)\chi_B(t))$$

as $c \le x \nearrow b^-$, for all $t \in \Omega$, and so

$$\mathcal{I}_\Phi(x, f\chi_P\chi_B) \nearrow I_{\widetilde{\Phi}}(f\chi_P\chi_B)$$

as $c \le x \nearrow b^-$, where

$$I_{\widetilde{\Phi}}(f) = \int_\Omega \widetilde{\Phi}(|f(t)|)\,d\mu(t),$$

for all $f \in L^0(\Omega)$. By the assumption on $a_w(x)$ applied to the function $g(x) = \mathcal{I}_\Phi(x, f\chi_P\chi_B)\chi_{[c,b[}(x)$, we obtain

$$\int_c^b a_w(x)\mathcal{I}_\Phi(x, f\chi_P\chi_B)\,dm(x) \to I_{\widetilde{\Phi}}(f\chi_P\chi_B), \quad \text{as } w \to w_0.$$

On the other hand, we have

$$\int_c^b a_w(x)\mathcal{I}_\Phi(x, f\chi_P\chi_B)\,dm(x) \le \int_c^b a_w(x)I_{\widetilde{\Phi}}(f\chi_P\chi_B)\,dm(x) \le I_{\widetilde{\Phi}}(f\chi_P\chi_B).$$

Hence

$$\begin{aligned} I_{\widetilde{\Phi}}(f\chi_P\chi_B) &= \lim_{w\to w_0}\int_c^b a_w(x)\mathcal{I}_\Phi(x, f\chi_P\chi_B)\,dm(x) \\ &= \sup_{w\in W}\int_c^b a_w(x)\mathcal{I}_\Phi(x, f\chi_P\chi_B)\,dm(x) \\ &= \Gamma_2(f\chi_P\chi_B). \end{aligned}$$

Taking $P = \Omega$ we obtain

$$I_{\widetilde{\Phi}}(f\chi_B) = \Gamma_2(f\chi_B) \le \mathcal{A}_\Phi(f) < +\infty.$$

Thus, there are a set $S_3 \in \Sigma$ with $\mu(S_3) < +\infty$ and a number $\delta_3 > 0$ such that

$$I_{\widetilde{\Phi}}(f\chi_{\Omega\setminus S_3}\chi_B) < \varepsilon/4,\, I_{\widetilde{\Phi}}(f\chi_S\chi_B) < \varepsilon/4, \quad \text{if } S\in\Sigma,\ \mu(S) < \delta_3.$$

Consequently,

$$\begin{aligned} &\Gamma_2(f\chi_{\Omega\setminus S_3}\chi_B) < \varepsilon/4 \\ &\Gamma_2(f\chi_S\chi_B) < \varepsilon/4, \quad \text{if } S\in\Sigma,\ \mu(S) < \delta_3. \end{aligned} \tag{2.2}$$

But we have for any set $P \in \Sigma$ and $x \in [c, b[$,

$$\mathcal{I}_\Phi(x, f\chi_P) = \mathcal{I}_\Phi(x, f\chi_P\chi_A) + \mathcal{I}_\Phi(x, f\chi_P\chi_B),$$

whence

$$\Gamma_2(f\chi_P) \le \Gamma_2(f\chi_P\chi_A) + \Gamma_2(f\chi_P\chi_B).$$

Consequently, by (2.1) and (2.2) we have, taking $S_4 = S_2 \cup S_3$, and $\delta_4 = \min\{\delta_2, \delta_3\}$,

$$\Gamma_2(f\chi_{\Omega\setminus S_4}) < \varepsilon/2, \quad \Gamma_2(f\chi_S) < \varepsilon/2,$$

with $S_4 \in \Sigma$, $\mu(S_4) < +\infty$, and $S \in \Sigma$, $\mu(S) < \delta_4$.

This shows that Γ_2 is a.c. We finally proved that $\mathcal{A}_\Phi$ is absolutely continuous. □

Remark 2.1. (a) Let us suppose that Ω is a locally compact and σ-compact Hausdorff topological group, equipped with its Haar measure μ. Then in the definition of absolute continuity of a monotone modular ρ on $L^0(\Omega)$ we can replace the set A in (a) with a *compact* A. Indeed, let ρ be monotone and absolutely continuous. In particular for every $\varepsilon > 0$ and $f \in L^0(\Omega)$ with $\rho(f) < +\infty$ there is a set $A \in \Sigma$, $A \subset \Omega$, such that $\mu(A) < +\infty$ and $\rho(\alpha f\chi_{\Omega\setminus A}) < \varepsilon/2$. Since Ω is σ-compact, we have $G = \bigcup_{k=1}^\infty W_n$, where W_n are compact (in the case this sum is finite, then Ω is compact itself and so we can take $A = \Omega$). Taking $V_n = \bigcup_{k=1}^n W_k$, we obtain again $\Omega = \bigcup_{n=1}^\infty V_n$, where V_n are compact, and $V_{n-1} \subset V_n$ for $n = 2, 3, \dots$. Hence

$$A = \bigcup_{n=1}^{\infty} A \cap V_n,$$

and $(A \cap V_n)$ is an increasing sequence of sets of finite measure in Σ. By a well-known theorem, we have

$$+\infty > \mu(A) = \lim_{n \to +\infty} \mu(A \cap V_n),$$

and so we obtain

$$\mu(A \setminus V_n) = \mu(A \setminus (A \cap V_n)) = \mu(A) - \mu(A \cap V_n) \to 0$$

as $n \to +\infty$. We now apply condition (b) of absolute continuity of ρ with $\varepsilon/2$ in place of ε. Let δ be a number corresponding to $\varepsilon/2$, and let $N \in \mathbb{N}$ be so large that for every $n > N$, $\mu(A \setminus V_n) \leq \delta$. Thus for $n > N$

$$\rho(\alpha f \chi_{A \setminus V_n}) < \varepsilon/2.$$

But since $\Omega \setminus V_n \subset (\Omega \setminus A) \cup (A \setminus V_n)$, we have $\chi_{\Omega \setminus V_n} \leq \chi_{\Omega \setminus A} + \chi_{A \setminus V_n}$. Hence for $n > N$

$$\begin{aligned} \rho\Big(\frac{\alpha}{2} f \chi_{\Omega \setminus V_n}\Big) &\leq \rho\Big(\frac{\alpha}{2} f \chi_{\Omega \setminus A} + \frac{\alpha}{2} f \chi_{A \setminus V_n}\Big) \\ &\leq \rho(\alpha f \chi_{\Omega \setminus A}) + \rho(\alpha f \chi_{A \setminus V_n}) \\ &< \varepsilon/2 + \varepsilon/2 = \varepsilon, \end{aligned}$$

and so the assertion follows.

(b) Let us still remark that additionally assuming the family of functions $\Phi(x, u)$ to be equicontinuous in $[a, b[$ at $u = 0$ we may show the modular $\mathcal{A}_\Phi$ to be absolutely finite. Indeed, for every $\varepsilon > 0$ there is a $\delta > 0$ such that, for any $u \in [0, \delta[$ and $x \in [a, b[$, we have $\Phi(x, u) < \varepsilon$. Now, let $A \in \Sigma$ and $\mu(A) < +\infty$. Then $\mathcal{I}_\Phi(x, \lambda \chi_A) = \mu(A)\Phi(x, \lambda)$, for every $\lambda > 0$. Hence

$$\mathcal{A}_\Phi(\lambda \chi_A) = \mu(A) \sup_{w \in W} \int_a^b a_w(x) \Phi(x, \lambda)\, dm(x) \leq \varepsilon \mu(A).$$

Consequently, $\mathcal{A}_\Phi$ is finite. The same inequality with $\varepsilon = 1$, $\lambda = \lambda_0$, $A = B$ shows that $\mathcal{A}_\Phi$ is absolutely finite.

Now, we are able to formulate the following modular version of the Lebesgue dominated convergence theorem.

Theorem 2.1. *Let ρ be a monotone, finite and absolutely continuous modular on a linear, correct subspace $\mathcal{X}$ of $L^0(\Omega)$. Let (f_n) be a sequence of functions $f_n \in \mathcal{X}$ such that $f_n(t) \to 0$ as $n \to +\infty$ μ-a.e. in Ω. Moreover, let us suppose there exists a function $g \in \mathcal{X}_\rho$ such that $\rho(3g) < +\infty$ and $|f_n(t)| \leq g(t)$ μ-a.e. in Ω, for $n = 1, 2, \ldots$ Then $\rho(f_n) \to 0$ as $n \to +\infty$.*

Proof. Let $\varepsilon > 0$ be arbitrary and let $\alpha > 0$, $A \in \Sigma$ and $\delta > 0$ be chosen as in the definition of absolute continuity, with f replaced by $3g$ and ε replaced by $\varepsilon/3$. By Egoroff's theorem, there exists a set $A_\varepsilon \in \Sigma$, $A_\varepsilon \subset A$, such that $\mu(A_\varepsilon) < \delta$ and $f_n(t) \to 0$ as $n \to +\infty$, uniformly on $A \setminus A_\varepsilon$. Since

$$|f_n(t)| = \frac{1}{3}[3|f_n(t)|\chi_{\Omega\setminus A}(t) + 3|f_n(t)|\chi_{A\setminus A_\varepsilon}(t) + 3|f_n(t)|\chi_{A_\varepsilon}(t)],$$

for $t \in \Omega$, applying monotonicity of ρ we obtain

$$\begin{aligned}\rho(\alpha f_n) &\le \rho(3\alpha f_n\chi_{\Omega\setminus A}) + \rho(3\alpha f_n\chi_{A\setminus A_\varepsilon}) + \rho(3\alpha f_n\chi_{A_\varepsilon})\\ &\le \rho(3\alpha g\chi_{\Omega\setminus A}) + \rho(3\alpha f_n\chi_{A\setminus A_\varepsilon}) + \rho(3\alpha g\chi_{A_\varepsilon}).\end{aligned}$$

By the choice of A_ε and δ we obtain

$$\rho(3\alpha g\chi_{\Omega\setminus A}) < \varepsilon/3 \quad \text{and} \quad \rho(3\alpha g\chi_{A_\varepsilon}) < \varepsilon/3.$$

Hence

$$\rho(\alpha f_n) \le \frac{2}{3}\varepsilon + \rho(3\alpha f_n\chi_{A\setminus A_\varepsilon}),$$

for $n = 1, 2, \ldots$ In view of the finiteness of ρ, we have $\chi_A \in \mathcal{X}_\rho$. Hence there exists a $\lambda_\varepsilon > 0$ such that $\rho(\lambda\alpha\chi_A) < \varepsilon/3$ for $0 < \lambda \le \lambda_\varepsilon$. Since $f_n(t) \to 0$ as $n \to +\infty$, uniformly in $A \setminus A_\varepsilon$, there exists an index n_0 such that $3\alpha|f_n(t)| < \lambda_\varepsilon$, for $t \in A \setminus A_\varepsilon$ and $n \ge n_0$. Hence, by monotonicity of ρ,

$$\rho(3\alpha f_n\chi_{A\setminus A_\varepsilon}) \le \rho(\lambda_\varepsilon\chi_{A\setminus A_\varepsilon}) \le \rho(\lambda_\varepsilon\chi_A) < \varepsilon/3,$$

for $n \ge n_0$. Consequently,

$$\rho(\alpha f_n) \le \frac{2}{3}\varepsilon + \rho(3\alpha f_n\chi_{A\setminus A_\varepsilon}) < \varepsilon,$$

for $n \ge n_0$. This proves that $\rho(\alpha f_n) \to 0$ as $n \to +\infty$. □

2.2 Moduli of continuity

One of the important tools in approximation theory and in other applications of mathematical analysis is the notion of modulus of continuity. We shall define it in the case of modular spaces $\mathcal{X}_\eta$ generated by a modular η in a linear correct subspace $\mathcal{X} \subset L^0(\Omega)$. Here, we assume Ω to be equipped with an operation $+ : \Omega \times \Omega \to \Omega$. For the sake of simplicity we shall suppose throughout this section that the operation $+$ is commutative. There is no problem to extend this to the case of a non- commutative operation and then the notions defined below, have their right-hand side and left-hand side versions (the reader may consult e.g. [117]).

In the following we shall need the notion of a filter $\mathcal{U}$ of subsets of Ω. We recall this notion: a family $\mathcal{U} \neq \emptyset$, of nonempty subsets of Ω is called a *filter* in Ω if it satisfies the following two conditions:

1. If $U_1, U_2 \in \mathcal{U}$, then $U_1 \cap U_2 \in \mathcal{U}$.
2. If $U_1 \in \mathcal{U}$, $U_2 \subset \Omega$, and $U_1 \subset U_2$ then $U_2 \in \mathcal{U}$.

Let us mention two important filters. In the first one, $\Omega = \mathbb{N}$ is the set of positive integers and the filter $\mathcal{U}$ of its subsets consists of complements of finite subsets of $\mathbb{N}$ (including the empty set). Convergence of a sequence $a_n \to a$, $(a_n, a \in \mathbb{R})$, as $n \to +\infty$ can be expressed saying that for every $\varepsilon > 0$ there exists a set $U_\varepsilon \in \mathcal{U}$ such that $|a_n - a| < \varepsilon$ for all $n \in U_\varepsilon$. In the second one, $\Omega = \mathbb{T}$ will mean a topological space and $\mathcal{U}$ is the family of all neighbourhoods of a fixed element $t_0 \in \mathbb{T}$. Again, a function $f : \Omega \to \mathbb{R}$ is convergent to $a \in \mathbb{R}$, if and only if for every $\varepsilon > 0$ there exists a set $U_\varepsilon \in \mathcal{U}$ such that $|f(t) - a| < \varepsilon$ for all $t \in U_\varepsilon$.

The notion of convergence may be generalized to a general filter $\mathcal{U}$ of subsets of Ω and obviously, we may restrict ourselves to convergence to zero. Namely, a function $f : \Omega \to \mathbb{R}$ is *$\mathcal{U}$-convergent* to zero if for every $\varepsilon > 0$ there is a set $U_\varepsilon \in \mathcal{U}$ such that $|f(t)| < \varepsilon$ for all $t \in U_\varepsilon$. We shall denote it writing $f(t) \xrightarrow{\mathcal{U}} 0$. One may define a *basis of a filter* as a family $\mathcal{U}_0 \subset \mathcal{U}$ such that for every set $U \in \mathcal{U}$ there exists a set $V \in \mathcal{U}_0$ such that $V \subset U$. Obviously, $f : \Omega \to \mathbb{R}$ is $\mathcal{U}$-convergent to zero, if and only if for every $\varepsilon > 0$ there exists a set $U_\varepsilon \in \mathcal{U}_0$ such that $|f(t)| < \varepsilon$, for all $t \in U_\varepsilon$. In the first of our previous examples of $\Omega = \mathbb{N}$ as $\mathcal{U}_0$ we may take the countable family of sets $U_n = \{n, n+1, n+2, \dots\}$, for $n = 1, 2, \dots$ In the second example of $\Omega = \mathbb{T}$, supposing $\mathbb{T}$ be a metric space and $t_0 \in \mathbb{T}$ to be fixed, we may take as a basis $\mathcal{U}_0$, the countable family of balls with centre at t_0 and radii $r_n = 1/n, n = 1, 2, \dots$

Applying the notion of a filter we shall specify a connection between the operation $+$ in Ω, the σ-algebra Σ and the measure μ in the measure space (Ω, Σ, μ). Let us denote for arbitrary $A \in \Omega$ and $t \in \Omega$,

$$A_t = \{s \in \Omega : t + s \in A,\ s \notin A,\ \text{or}\ t + s \notin A,\ s \in A\}.$$

Let $\mathcal{U}$ be a filter in Ω. We say that $\{\Omega, \mathcal{U}, \Sigma, \mu\}$ is a *correctly filtered system* with respect to $\mathcal{X}$, if

1. the filter $\mathcal{U}$ contains a basis $\mathcal{U}_0 \subset \Sigma$,
2. if $A \in \Sigma$ and $\mu(A) < +\infty$, then $A_t \in \Sigma$ for every $t \in \Omega$ and $\mu(A_t) \xrightarrow{\mathcal{U}} 0$,
3. $\mathcal{X}$ is invariant with respect to the operation $+$.

Example 2.3. Let $(\Omega, +)$ be an abelian locally compact Hausdorff topological group. For $\mathcal{U}$ we take the filter of neighbourhoods of the neutral element of Ω and μ will denote the Haar measure on $(\Omega, +)$. Denoting by $A \Delta B$ the symmetric difference of sets A and B, we have $A_t = A\Delta(A - t)$ for each $A \subset \Omega$ and $t \in \Omega$. It is well-known that if $A \in \Sigma$ then $A_t \in \Sigma$ for each $t \in \Omega$ and $\mu(A_t) \xrightarrow{\mathcal{U}} 0$ (see e.g. [117]). Moreover, if f is Σ-measurable and $a \in \mathbb{R}$, $t \in \Omega$, we have

$$\{s \in \Omega : f(t + s) > a\} = \{\sigma - t \in \Omega : f(\sigma) > a\} = \{\sigma \in \Omega : f(\sigma) > a\} - t \in \Sigma,$$

whence $f(t+\cdot)$ is also Σ-measurable. Thus, the system $\{\Omega, \mathcal{U}, \Sigma, \mu\}$ is correctly filtered.

Let $\mathcal{U}$ be a filter on Ω with basis $\mathcal{U}_0 \in \Sigma$ and let $\mathcal{X}$ be an invariant, correct linear subspace of $L^0(\Omega)$. Let η be a modular on $\mathcal{X}$. The η-*modulus of continuity* is defined as the map

$$\omega_\eta : \mathcal{X} \times \mathcal{U} \to \widetilde{\mathbb{R}}_0^+,$$

where

$$\omega_\eta(f, U) = \sup_{t\in U} \eta(f(t+\cdot) - f(\cdot)),$$

for all $f \in \mathcal{X}$ and $U \in \mathcal{U}$. The elementary properties of a modulus of continuity are summarized in the following

Theorem 2.2. *If η is a monotone modular on $\mathcal{X}$, then*

(a) $\omega_\eta(f, V) \le \omega_\eta(f, U)$, *for* $f \in \mathcal{X}, U, V \in \mathcal{U}, V \subset U$,

(b) $\omega_\eta(|f|, U) \le \omega_\eta(f, U)$, *for* $f \in \mathcal{X}, U \in \mathcal{U}$,

(c) $\omega_\eta(af, U) \le \omega_\eta(bf, U)$, *for* $f \in \mathcal{X}, U \in \mathcal{U}, 0 \le a \le b$,

(d) $\omega_\eta(\sum_{j=1}^n f_j, U) \le \sum_{j=1}^n \omega_\eta(nf_j, U)$, *for* $f_1, f_2, \dots f_n \in \mathcal{X}, U \in \mathcal{U}$.

Proof. Properties (a), (c), (d) are obvious. Applying the fact that $\mathcal{X}$ is a correct subspace of $L^0(\Omega)$ so that $f \in \mathcal{X}$ implies $|f| \in \mathcal{X}$, and the monotonicity of η, we obtain

$$\eta(|f(t+\cdot)| - |f(\cdot)|) \le \eta(|f(t+\cdot) - f(\cdot)|) = \eta(f(t+\cdot) - f(\cdot)),$$

which implies (b). □

We now solve the problem, under what assumptions the η-modulus of continuity of a function $f \in \mathcal{X}$ tends to zero in the sense of the filter $\mathcal{U}$. First, we consider the case when the function f is a simple function, vanishing outside a set of finite measure μ. If $\mathcal{X}$ is a correct linear subspace of $L^0(\Omega)$ then all such functions belong to $\mathcal{X}$.

Theorem 2.3. *Let $(\Omega, \mathcal{U}, \Sigma, \mu)$ be a correctly filtered system and let $\mathcal{X}$ be an invariant, correct linear subspace of $L^0(\Omega)$. Let η be a monotone, absolutely finite modular on $\mathcal{X}$. Then for any simple function f on Ω, vanishing outside a set of finite measure μ and for every $\lambda > 0$ there holds the relation $\omega_\eta(\lambda f, U) \xrightarrow{\mathcal{U}} 0$.*

Proof. Let $f = \chi_A$ be the characteristic function of a set $A \in \Sigma$ of measure $\mu(A) < +\infty$. It is easily verified that

$$|\chi_A(t+s) - \chi_A(s)| = \chi_{A_t}(s),$$

for every $s, t \in \Omega$. Since $\mu(A) < +\infty$ and $(\Omega, \mathcal{U}, \Sigma, \mu)$ is correctly filtered, so $A_t \in \Sigma$, for $t \in \Omega$ and $\mu(A_t) \xrightarrow{\mathcal{U}} 0$. Thus, there exists a $U_0 \in \mathcal{U}$ such that $\mu(A_t) < +\infty$ for $t \in U_0$. Since η is finite, we have $\chi_A \in \mathcal{X}_\eta$ and $\chi_{A_t} \in \mathcal{X}_\eta$ for $t \in U_0$. Since $\mathcal{X}_\eta$ is linear, we thus have $\lambda_0 \chi_{A_t} \in \mathcal{X}_\eta$, for $t \in U_0$ and every $\lambda_0 > 0$. Let $\varepsilon > 0$ be arbitrary. Since η is absolutely finite, there exists a $\delta > 0$ (depending on ε and λ_0), such that if $B \in \Sigma$ and $\mu(B) < \delta$, then $\eta(\lambda_0 \chi_B) < \varepsilon$. Since $\mu(A_t) \xrightarrow{\mathcal{U}} 0$, there is a set $U \in \mathcal{U}$, $U \subset U_0$ such that $\mu(A_t) < \delta$ for all $t \in U$. Consequently, $\eta(\lambda_0 \chi_{A_t}) < \varepsilon$ for all $t \in U$. Thus, $\sup_{t \in U} \eta(\lambda_0 \chi_{A_t}) = \omega_\eta(\lambda_0 \chi_A, U) \leq \varepsilon$. This shows that $\omega_\eta(\lambda_0 \chi_A, U) \xrightarrow{\mathcal{U}} 0$.

Now, let $f = \sum_{j=1}^{n} c_j \chi_{A_j}$, where $A_j \in \Sigma$, $\mu(A_j) < +\infty$ for $j = 1, 2, \ldots, n$ and $A_1, A_2, \ldots, A_n$ are pairwise disjoint. Since $\chi_{A_j} \in \mathcal{X}_\eta$, for $j = 1, 2, \ldots, n$, so $f \in \mathcal{X}_\eta$. Let $\lambda > 0$ be arbitrary and let $\lambda_0 = n\lambda \max_j |c_j|$. Applying Theorem 2.2 (d) and (c), we obtain the inequality

$$\omega_\eta(\lambda f, U) \leq \sum_{j=1}^{n} \omega_\eta(\lambda_0 \chi_{A_j}, U).$$

From the first part of the proof we conclude that $\omega_\eta(\lambda_0 \chi_{A_j}, U) \xrightarrow{\mathcal{U}} 0$, for $j = 1, 2, \ldots, n$. Consequently, $\omega_\eta(\lambda f, U) \xrightarrow{\mathcal{U}} 0$. □

In Section 1.4 we introduced the notion of a subbounded modular and a strongly subbounded modular with respect to the operation $+$. Applying the filter $\mathcal{U}$ we distinguish now a notion between the two above ones. Namely a modular η in a linear subspace $\mathcal{X}$ of $L^0(\Omega)$ will be called *bounded* (with respect to the operation $+$ and a filter $\mathcal{U}$ in Ω), if there are a constant $C \geq 1$ and a function $\ell : \Omega \to \mathbb{R}_0^+$ satisfying the conditions $\ell \in L^\infty(\Omega)$, $\ell(t) \xrightarrow{\mathcal{U}} 0$, such that for every function $f \in \mathcal{X}$ and every $t \in \Omega$ there holds the inequality

$$\eta(f(t + \cdot)) \leq \eta(Cf) + \ell(t).$$

Obviously, a strongly subbounded modular η is always bounded, and a bounded modular η is always subbounded.

Example 2.4. Let the system $\{\Omega, \mathcal{U}, \Sigma, \mu\}$ be defined as in Example 2.3 and let the function φ be as in Example 1.5 (c). Moreover, let φ satisfy the inequality (1.9) in Example 1.10 (b), for $s, t \in \Omega$, $u \geq 0$, where $C \geq 1$ is a constant and $0 \leq h(\cdot, t) \in L^1(\Omega)$, $\int_\Omega h(s, t)\, d\mu(s) \to 0$ as $t \to \theta$, where θ is the neutral element of the group Ω. Following the estimates in Example 1.10 (b) one may check easily that the modular $\eta(f) = \int_\Omega \varphi(t, |f(t)|)\, d\mu(t)$ is bounded with respect to the operation $+$ and the filter $\mathcal{U}$ of neighbourhoods of θ in Ω.

Now we may generalize Theorem 2.3 to the whole space $\mathcal{X}$.

Theorem 2.4. *Let $\{\Omega, \mathcal{U}, \Sigma, \mu\}$ be a correctly filtered system and let $\mathcal{X}$ be an invariant, correct linear subspace of $L^0(\Omega)$. Let η be a monotone, absolutely finite, absolutely continuous and bounded modular on $\mathcal{X}$. Then for every function $f \in \mathcal{X}_\eta$ there exists a number $\lambda > 0$ such that*

$$\omega_\eta(\lambda f, U) \xrightarrow{\mathcal{U}} 0.$$

Proof. Let us first remark that it is sufficient to prove the theorem for functions $f \geq 0$. Indeed, suppose the theorem to be true for functions $f \in \mathcal{X}_\eta$, $f \geq 0$ and let $f \in \mathcal{X}_\eta$, be arbitrary. Denoting by f_+, f_- the positive part and the negative part of f, we have $f_+ = (1/2)(|f| + f)$, $f_- = (1/2)(|f| - f)$. Since $\mathcal{X}$ is a correct subspace, the assumption $f \in \mathcal{X}$ implies $|f| \in \mathcal{X}$, and since η is monotone, so $\eta(f) = \eta(|f|)$, and from $f \in \mathcal{X}_\eta$, we conclude that $|f| \in \mathcal{X}_\eta$. Since $\mathcal{X}_\eta$ is linear, this implies $f_+, f_- \in \mathcal{X}_\eta$. Thus, by Theorem 2.2 (d) we obtain

$$\begin{aligned}\omega_\eta\left(\frac{1}{2}\lambda f, U\right) &= \omega_\eta\left(\frac{1}{2}\lambda(f_+ - f_-), U\right) \\ &\leq \omega_\eta(\lambda f_+, U) + \omega_\eta(\lambda f_-, U) \xrightarrow{\mathcal{U}} 0,\end{aligned}$$

since $f_+, f_- \geq 0$. Hence we may restrict the proof to functions $f \geq 0$.

Let $f \in \mathcal{X}_\eta$, $f \geq 0$. There exists a sequence (g_n) of nonnegative simple functions such that $g_n(t) \nearrow f(t)$ as $n \nearrow +\infty$, μ-a.e. in Ω. Since the measure μ is σ-finite, we may define g_n in such a manner that each g_n vanishes outside a set $A_n \in \Sigma$ of finite measure μ. Since the modular η is finite, we have $g_n \in \mathcal{X}_\eta$, for $n = 1, 2, \ldots$. Hence also $f_n = f - g_n \in \mathcal{X}_\eta$, for $n = 1, 2, \ldots$. Moreover, $0 \leq f_n(t) \searrow 0$ as $n \nearrow +\infty$ and $f_n(t) \leq f(t)$ for all $t \in \Omega$. Since $f \in \mathcal{X}_\eta$, there exists a number $\lambda_0 > 0$ such that $\eta(3\lambda_0 f) < +\infty$. Applying Theorem 2.1 with $g(t) = \lambda_0 f(t)$, we obtain $\eta(\lambda_0 f_n) \to 0$ as $n \to +\infty$. Additionally, we may take λ_0 so small that $\eta(\lambda_0 f) < +\infty$. Selecting eventually a subsequence from (g_n) we may suppose that

$$\eta(\lambda_0(f - g_n)) < \frac{1}{n}, \qquad n = 1, 2, \ldots \tag{2.3}$$

Let $C \geq 1$ and $\ell \in L^\infty(\Omega)$, $\ell(t) \xrightarrow{\mathcal{U}} 0$, be as in the definition of a bounded modular. Since $\eta(\lambda_0(f - g_n)) < +\infty$, for $n = 1, 2, \ldots$, we may apply the definition of boundedness of η to the function $(\lambda_0/C)(f - g_n)$ in place of f, obtaining the inequality

$$\eta\left(\frac{\lambda_0}{C}(f(t + \cdot) - g_n(t + \cdot))\right) \leq \eta(\lambda_0(f - g_n)) + \ell(t), \qquad \text{for } t \in \Omega.$$

This implies

$$\begin{aligned}\eta\Big(\frac{\lambda_0}{3C}(f(t+\cdot)-f(\cdot))\Big) &\le \eta\Big(\frac{\lambda_0}{C}(f(t+\cdot)-g_n(t+\cdot))\Big)\\ &\quad+\eta\Big(\frac{\lambda_0}{C}(g_n(t+\cdot)-g_n(\cdot))\Big)+\eta\Big(\frac{\lambda_0}{C}(g_n-f)\Big)\\ &\le \frac{2}{n}+\ell(t)+\omega_\eta(\lambda_0 g_n, U),\end{aligned}$$

for arbitrary $U\in\mathcal{U}, t\in U, n=1,2,\ldots$. Let us take an arbitrary $\varepsilon>0$ and let us fix an index n in such a manner that $2/n\le\varepsilon/3$. By Theorem 2.3, there exists a $U_1\in\mathcal{U}$ such that for every $V\in\mathcal{U}$, $V\subset U_1$ there holds $\omega_\eta(\lambda_0 g_n, V)<\varepsilon/3$. Since $\ell(t)\xrightarrow{\mathcal{U}}0$, there is a $U_2\in\mathcal{U}$ such that $\ell(t)<\varepsilon/3$ for $t\in U_2$. Consequently, taking $t\in U_1\cap U_2\in\mathcal{U}$, we obtain

$$\eta\Big(\frac{\lambda_0}{3C}(f(t+\cdot)-f(\cdot))\Big)\le\varepsilon.$$

Thus, $\omega_\eta((\lambda_0/3C)f, V)\le\varepsilon$, for $V\in\mathcal{U}$, $V\subset U_1\cap U_2$. This shows that

$$\omega_\eta((\lambda_0/3C)f, U)\xrightarrow{\mathcal{U}}0. \qquad \square$$

Example 2.5. (a) Let $\Omega=\mathbb{R}_0^+$ be provided with the operation of usual addition $+$ and let μ be the Lebesgue measure in the σ-algebra of all Lebesgue measurable subsets on $\mathbb{R}_0^+$. Let φ be the function from Example 1.5 (c) and let

$$\eta(f)=I_\varphi(f)=\int_0^{+\infty}\varphi(t,|f(t)|)\,dt.$$

Let $\mathcal{U}$ be a filter in $\mathbb{R}_0^+$ with basis $\mathcal{U}_0$ consisting of all intervals of the form $U_\delta=[0,\delta[$ with $\delta>0$. Obviously, $\{\mathbb{R}_0^+,\mathcal{U},\Sigma,\mu\}$ is a correctly filtered system. The space $L_\eta^0(\mathbb{R}_0^+)$ is equal to the generalized Orlicz space $L^\varphi(\mathbb{R}_0^+)$, and if we restrict the modulus ω_η to the basis $\mathcal{U}_0$, denoting $\omega_\varphi(f,\delta)=\omega_\eta(f,U_\delta)$, we obtain, by Theorem 2.4 that for every $f\in L^\varphi(\mathbb{R}_0^+)$, there exists a number $\lambda>0$ such that

$$\omega_\varphi(\lambda f,\delta)=\sup_{0<t<\delta}\int_0^{+\infty}\varphi(s,\lambda|f(t+s)-f(s)|)\,ds\to 0,\quad \text{as } \delta\to 0^+.$$

(b) Let $\Omega=]0,1]$ provided with the operation of usual multiplication $\cdot$ and let μ be the measure defined by $d\mu(t)=dt/t$, where dt is the Lebesgue measure in the σ-algebra of all Lebesgue measurable subsets on $]0,1]$. Let $\varphi(t,u)$ be the function from Example 1.5 (c), defined for $(t,u)\in]0,1]\times\mathbb{R}$, and eventually extended by 1-periodicity with respect to the first variable, to the whole $\mathbb{R}^+$. Let

$$\eta(f)=I_\varphi(f)=\int_0^1\varphi(t,|f(t)|)\,\frac{dt}{t}.$$

Let $\mathcal{U}$ be a filter in $]0,1]$ with basis $\mathcal{U}_0$ consisting of all intervals of the form $U_\delta =]1-\delta, 1]$, for $\delta \in]0, 1/2[$. Then $L^0_\eta(\Omega)$ is the space of all measurable functions $f : \Omega \to \mathbb{R}$, such that there is $\lambda > 0$ for which

$$\int_0^1 \varphi(t, \lambda|f(t)|)\,\frac{dt}{t} < +\infty.$$

As before, it is easily seen that $\{]0,1], \mathcal{U}, \Sigma, \mu\}$ is a correctly filtered system, and putting again $\omega_\varphi(f,\delta) = \omega_\eta(f, U_\delta)$, we obtain

$$\omega_\varphi(\lambda f, \delta) = \sup_{1-\delta<t<1} \int_0^1 \varphi(s, \lambda|f(ts) - f(s)|)\,ds \to 0, \qquad \text{as } \delta \to 0^+.$$

(c) Let $\Omega = \mathbb{N}_0 = \{0,1,2,\dots\}$ be the semigroup of nonnegative integers, endowed with the usual operation of addition $+$ and with the counting measure μ on the σ-algebra of all subsets of Ω. As before, we take into consideration the filter $\mathcal{U}$ consisting of the complements of all finite subsets of Ω with basis $\mathcal{U}_0$ given by the sets $\{n, n+1, n+2, \dots\}$, $n = 1, 2, \dots$. Then again $\{\mathbb{N}, \mathcal{U}, \Sigma, \mu\}$ is a correctly filtered system. Let η be any modular defined on the space $\ell^0(\Omega)$, consisting of all the sequences (t_j) of real numbers, satisfying all the assumptions of Theorem 2.4. Then for a sequence $f = (t_j)_{j\in\mathbb{N}_0}$, in the corresponding modular space, denoted by ℓ^0_η, there holds

$$\omega_\eta(\lambda f, U) = \sup_{j\ge n} \eta(\lambda(t_{\cdot+j} - t_\cdot)) \xrightarrow{\mathcal{U}} 0.$$

2.3 Bibliographical notes

The notions of a monotone modular and a finite modular were introduced in [155], as a generalization of classical Köthe norms (see [139]). The concepts of absolutely finite and absolutely continuous modular were defined in [20] (see also [155]). The notion of a correct subspace $\mathcal{X}$ of $L^0(\Omega)$ may also be found in [20]. Spaces connected with strong summability were first considered in [148] (general matrix methods) and [165] (first arithmetic means), and then the investigations were continued by A. Waszak [203]. For Example 2.2 (b), see [18]. Theorem 2.1 plays a key role in applications of modular spaces and was proved in [155].

In order to define a modulus of continuity generated by means of a modular η, we need an operation $+$ from $\Omega \times \Omega$ to Ω. We do not need to suppose $(\Omega, +)$ to be a group. However, we need some continuity property of $+$ which leads to the notion of a correctly filtered system $\{\Omega, \mathcal{U}, \Sigma, \mu\}$, where $\mathcal{U}$ is a filter of subsets of Ω with a basis $\mathcal{U}_0$ of Σ-measurable subsets. This notion was introduced in [20] and further applied in [22]. The connection between this notion and the special case of a locally compact topological group structure is explained by Example 2.3.

The notion of a modulus of continuity of a function in various function spaces, especially in Banach function spaces, belongs to fundamental tools in the theory of approximation (see e.g. [67]). It was transferred to the more general case of modular spaces in [155], where its basic properties were also investigated. The proof of Theorem 2.4, a fundamental property stating that the η-modulus of a function in a modular space $L^0_\eta(\Omega)$ tends to zero in the sense of the η-convergence was finally proved in [20]. It requires a notion of a bounded modular η introduced in [155] and then applied in [19] and in other papers, where it was termed as τ-bounded modular.

Chapter 3

Approximation by convolution type operators

3.1 Embedding theorems and the error of modular approximation

In this chapter we shall deal with *convolution-type operators* of the form

$$(Tf)(s) = \int_{\Omega} K(t, f(t+s))\, d\mu(t), \tag{3.1}$$

defined by Example 1.2. Here, (Ω, Σ, μ) will be a measure space with a σ-finite, complete measure, $+$ is a commutative operation from $\Omega \times \Omega$ to Ω. We will assume that this operation is a measurable function from $\Omega \times \Omega$ to Ω. The function $K : \Omega \times \mathbb{R} \to \mathbb{R}$ is a *Carathéodory kernel function*, i.e. it is Σ-measurable in Ω for every $u \in \mathbb{R}$, and it is continuous in $\mathbb{R}$ for every $t \in \Omega$, with $K(t, 0) = 0$. Let $L^0(\Omega)$ be the space of all extended real-valued, Σ-measurable and finite μ-a.e. functions f. It is well-known that if $f \in L^0(\Omega)$, then $K(t, f(t+s))$ is a Σ-measurable function of $t \in \Omega$, for every $s \in \Omega$. As in 1.1, the *domain* Dom T of the operator T defined by (3.1) is defined as the set of all functions $f \in L^0(\Omega)$ for which the integral (3.1) exists for a.e. $s \in \Omega$ and $(Tf)(s)$ is a Σ-measurable function of $s \in \Omega$. A special case of the kernel function K is obtained when we suppose $K(t, u)$ to be a linear function of u, i.e. $K(t, u) = \widetilde{K}(t)u$, where $\widetilde{K} : \Omega \to \mathbb{R}$ is Σ-measurable on Ω. The convolution-type operator (3.1) takes on the form

$$(Tf)(s) = \int_{\Omega} \widetilde{K}(t) f(t+s)\, d\mu(t).$$

Operators of this form are used in approximation theory since the beginnings of this theory.

Example 3.1. (a) Let $f \in L^1_{2\pi}$, i.e. f is a 2π-periodic, real valued function, Lebesgue integrable in the interval $[-\pi, \pi]$. Let S_n be the n-th partial sum of the Fourier series of the function f, then

$$S_n(s) = \int_{-\pi}^{\pi} D_n(t) f(t+s)\, dt,$$

for $n \in \mathbb{N}$, where

$$D_n(u) = \frac{1}{2\pi} \frac{\sin[(2n+1)(u/2)]}{\sin(u/2)}, \quad \text{for } 0 < |u| \leq \pi$$

are the *Dirichlet kernel functions.*

(b) Let (σ_n) be the sequence of arithmetic means of the sequence (S_n) from (a), then

$$\sigma_n(s) = \frac{1}{n+1} \sum_{k=0}^{n} S_k(s) = \int_{-\pi}^{\pi} K_n(t) f(t+s)\, ds$$

for $n \in \mathbb{N}$, where

$$K_n(u) = \frac{1}{2\pi(n+1)} \left(\frac{\sin[(n+1)u/2]}{\sin(u/2)} \right)^2 \quad \text{for } 0 \le |u| \le \pi$$

are the *Fejér kernel functions.*

(c) Let $f \in L^1_{2\pi}$ and let

$$\mathcal{A}_r(s) = \frac{1}{2} a_0 + \sum_{n=1}^{\infty} r^n (a_n \cos ns + b_n \sin ns),$$

where (a_n), (b_n) are the sequences of Fourier coefficients of the function f. Since the sequences (a_n) and (b_n) are bounded, the series $\mathcal{A}_r(s)$ is uniformly convergent for $s \in [-\pi, \pi]$, if $r \in [0, 1[$ is fixed. Moreover, we have

$$\mathcal{A}_r(s) = \int_{-\pi}^{\pi} A_r(t) f(t+s)\, ds$$

for $r \in [0, 1[$, where

$$A_r(t) = \frac{1}{2\pi} \frac{1 - r^2}{1 - 2r \cos t + r^2} \quad \text{for } |t| \le \pi$$

are the *Abel–Poisson kernel functions.*

One of the fundamental questions in the theory of Fourier series is, under what conditions the above defined sequences of operators (S_n), (σ_n) and $(\mathcal{A}_r)$ tend to f as $n \to +\infty$ in the first two cases and as $r \to 1^-$ in the third one. This also depends on the kind of convergence we require: almost everywhere, pointwise, uniform (i.e. in the space $C_{2\pi}$ of 2π-periodic, continuous functions), in $L^1_{2\pi}$, etc.

In case of operators (3.1), the linear operators are replaced by nonlinear ones. In order to apply operators (3.1) in approximation theory, we first have to find a tool, which would replace linearity of a kernel function. This will be a generalized Lipschitz condition. Let $L : \Omega \to \mathbb{R}_0^+ = [0, +\infty[$ be Σ-measurable, $0 \ne L \in L^1(\Omega)$, and let us put $\|L\|_1 = \int_\Omega L(t)\, d\mu(t)$, $p(t) = L(t)/\|L\|_1$, for $t \in \Omega$; obviously, $\|p\|_1 = 1$. Let Ψ be the class of all functions $\psi : \Omega \times \mathbb{R}_0^+ \to \mathbb{R}_0^+$ which satisfy the assumptions given in 1.5, i.e. $\psi(\cdot, u)$ is Σ-measurable for every $u \ge 0$, $\psi(t, :)$ is continuous and nondecreasing, for every $t \in \Omega$, $\psi(t, 0) = 0$, $\psi(t, u) > 0$, for $u > 0$, $\psi(t, u) \to +\infty$

as $u \to +\infty$, for all $t \in \Omega$. We say that the kernel function is *(L, ψ)-Lipschitz*, briefly $K \in (L, \psi)$-Lip, if for all $t \in \Omega$, $u, v \in \mathbb{R}$, there holds the following inequality

$$|K(t,u) - K(t,v)| \le L(t)\psi(t, |u-v|). \tag{3.2}$$

Moreover, we say that K is *$(L, \psi)_0$-Lipschitz*, briefly $K \in (L, \psi)_0$-Lip, if for all $t \in \Omega$, $u \in \mathbb{R}$ there holds the inequality

$$|K(t,u)| \le L(t)\psi(t, |u|). \tag{3.3}$$

Since $K(t, 0) = 0$, if K is (L, ψ)-Lipschitz, it is also $(L, \psi)_0$-Lipschitz.

In case of a linear kernel $K(t, u) = \widetilde{K}(t)u$, an important role in approximation theory play the assumptions of singularity of the function $\widetilde{K}$. For example, one of the fundamental tools in approximating 2π-periodic functions f, integrable in the interval $[-\pi, \pi]$, by means of the sequence (σ_n) of first arithmetic means of partial sums of its Fourier series (Example 3.1 (b)), is the singularity of these kernel functions, defined by the conditions

$$\int_{-\pi}^{-\delta} \widetilde{K}_n(t)\,dt \to 0, \quad \text{and} \quad \int_{\delta}^{\pi} \widetilde{K}_n(t)\,dt \to 0,$$

as $n \to +\infty$, for any $\delta \in]0, \pi[$ and

$$\int_{-\pi}^{\pi} \widetilde{K}_n(t)\,dt = 1, \quad \text{for } n = 1, 2, \ldots$$

It is obvious that, in analogy to the linear case, also in the general case of convolution-type operators one will need some singularity assumptions on the kernel functions in order to obtain results in approximation theory.

Finally, one should decide about the notion of convergence used for the approximation of f by means of Tf. We shall apply here modular convergence in the space $L^0_\rho(\Omega)$, generated by a modular ρ, i.e. we are going to estimate the error of approximation $\rho(\alpha(Tf - f))$. Due to Example 1.5 (a), this includes approximation in the sense of a norm in a normed linear subspace of the space $L^0(\Omega)$. However, first we shall investigate the problem of continuity of the operator T from a modular space L^0_η generated by a modular η to a modular space $L^0_\rho(\Omega)$ generated by a modular ρ.

We prove the following

Theorem 3.1. *Let ρ be a monotone, quasiconvex with a constant $M \ge 1$ modular on $L^0(\Omega)$ and let η be a modular on $L^0(\Omega)$, subbounded with respect to the operation $+$ with constant $C \ge 1$ and function $\ell \in L^\infty(\Omega)$. Let K be an $(L, \psi)_0$-Lipschitz Carathéodory kernel function such that $\{\rho, \psi, \eta\}$ is a properly directed triple and let the operator T be defined by* (3.1). *Finally, let $U \in \Sigma$, $0 < \lambda < 1$ and $0 < \alpha < C_\lambda(M\|L\|_1)^{-1}$ be arbitrary. Then for every $f \in L^0_\eta(\Omega) \cap \operatorname{Dom} T$ there holds the inequality*

$$\rho(\alpha Tf) \le M\eta(C\lambda f) + M\|\ell\|_\infty \int_{\Omega\setminus U} p(t)\,d\mu(t) + M \operatorname{ess\,sup}_{t\in U} \ell(t).$$

Consequently, $T : L^0_\eta(\Omega) \cap \operatorname{Dom} T \to L^0_\rho(\Omega)$.

Proof. Since $K \in (L, \psi)_0$-Lip, we have

$$|Tf(s)| \leq \int_\Omega p(t)\|L\|_1 \psi(t, |f(t+s)|)\, d\mu(t).$$

Applying monotonicity and quasiconvexity of ρ and the assumption that $\{\rho, \psi, \eta\}$ is a properly directed triple, we obtain for $0 < \lambda < 1$ and $\alpha > 0$ such that $M\|L\|_1\alpha \leq C_\lambda$ the inequalities

$$\begin{aligned}
\rho(\alpha Tf) &\leq \rho\left(\alpha \int_\Omega p(t)\|L\|_1 \psi(t, |f(t+\cdot)|)\, d\mu(t)\right) \\
&\leq M \int_\Omega p(t)\rho[M\|L\|_1 \alpha\psi(t, |f(t+\cdot)|]\, d\mu(t) \\
&\leq M \int_\Omega p(t)\eta(\lambda |f(t+\cdot)|)\, d\mu(t).
\end{aligned}$$

Since η is subbounded, we have $\eta(\lambda|f(t+\cdot)|) \leq \eta(C\lambda f) + \ell(t)$, where $\|\ell\|_\infty < +\infty$. Hence

$$\begin{aligned}
\rho(\alpha Tf) &\leq M \int_\Omega p(t)\eta(C\lambda f)\, d\mu(t) + M \int_\Omega p(t)\ell(t)\, d\mu(t) \\
&\leq M\eta(C\lambda f) + M \int_{\Omega \setminus U} p(t)\ell(t)\, d\mu(t) + M \int_U p(t)\ell(t)\, d\mu(t) \\
&\leq M\eta(C\lambda f) + M\|\ell\|_\infty \int_{\Omega \setminus U} p(t)\, d\mu(t) + M \operatorname{ess\,sup}_{t \in U} \ell(t)
\end{aligned}$$

for arbitrary $U \in \Sigma$. □

Now, as an immediate consequence, we may state the following

Corollary 3.1. *Let ρ be a monotone, quasiconvex with a constant $M \geq 1$ modular on $L^0(\Omega)$ and let η be a modular on $L^0(\Omega)$, strongly subbounded with respect to the operation $+$ with constant $C \geq 1$. Let K be an $(L, \psi)_0$-Lipschitz Carathéodory kernel function such that $\{\rho, \psi, \eta\}$ is a properly directed triple and let the operator T be defined by* (3.1). *Let $0 < \lambda < 1$ and $0 < \alpha < C_\lambda(M\|L\|_1)^{-1}$ be arbitrary. Then for every $f \in L^0_\eta(\Omega) \cap \operatorname{Dom} T$ there holds the inequality*

$$\rho(\alpha Tf) \leq M\eta(C\lambda f)$$

and consequently the operator T is continuous at 0 *in the sense of modular convergence.*

The last part of Corollary 3.1 means that if $f_n \in L^0_\eta(\Omega) \cap \mathrm{Dom}\, T$ for $n = 1, 2, \ldots$, then $f_n \xrightarrow{\eta} 0$ implies $Tf_n \xrightarrow{\rho} 0$.

The next theorem will give an estimation of the error of modular approximation $\rho(\alpha(Tf - f))$.

Theorem 3.2. *Let ρ be a monotone, J-quasiconvex with a constant $M \geq 1$ modular on $L^0(\Omega)$ and let η be a modular on $L^0(\Omega)$, subbounded with respect to the operation $+$ with a constant $C \geq 1$ and a function $\ell \in L^\infty(\Omega)$. Let K be an (L, ψ)-Lipschitz Carathéodory kernel function such that $\{\rho, \psi, \eta\}$ is a properly directed triple and let the operator T be defined by (3.1). Finally, let $U \in \Sigma$, $0 < \lambda < 1$ and $0 < \alpha < C_\lambda(2M\|L\|_1)^{-1}$ be arbitrary. Then for every $f \in L^0_\eta(\Omega) \cap \mathrm{Dom}\, T$ there holds the inequality*

$$\rho[\alpha(Tf - f)] \leq M\omega_\eta(\lambda f, U) + M[2\eta(2C\lambda f) + \|\ell\|_\infty] \int_{\Omega \setminus U} p(t)\, d\mu(t) + R, \tag{3.4}$$

where

$$R = \rho\left(2\alpha \left| \int_\Omega K(t, f(\cdot))\, d\mu(t) - f(\cdot) \right| \right). \tag{3.5}$$

Proof. Since

$$Tf(s) - f(s) = \int_\Omega [K(t, f(t+s)) - K(t, f(s))]\, d\mu(t) + \int_\Omega K(t, f(s))\, d\mu(t) - f(s)$$

so, by the (L, ψ)-Lipschitz condition, we obtain

$$|Tf(s) - f(s)| \leq \int_\Omega p(t)\|L\|_1 \psi(t, |f(t+s) - f(s)|)\, d\mu(t) + \left| \int_\Omega K(t, f(s))\, d\mu(t) - f(s) \right|$$

for $s \in \Omega$. Since ρ is a monotone modular, we get

$$\rho[\alpha(Tf - f)] \leq \rho\left[2\alpha \int_\Omega p(t)\|L\|_1 \psi(t, |f(t + \cdot) - f(\cdot)|)\, d\mu(t)\right] + \rho\left[2\alpha \left| \int_\Omega K(t, f(\cdot))\, d\mu(t) - f(\cdot) \right| \right].$$

Denoting by J_1 the first term at the right-hand side of the last inequality, we have

$$\rho[\alpha(Tf - f)] \leq J_1 + R. \tag{3.6}$$

By quasiconvexity of ρ, we obtain

$$J_1 \leq M \int_\Omega p(t)\rho[2\alpha M\|L\|_1 \psi(t, |f(t+\cdot) - f(\cdot)|)]\, d\mu(t).$$

Since $\{\rho, \psi, \eta\}$ is a properly directed triple, so choosing, for a given $\lambda \in]0,1[$, a number α such that $0 < 2\alpha M\|L\|_1 \leq C_\lambda$ we obtain, for an arbitrary $U \in \Sigma$, the inequality

$$\begin{aligned} J_1 \leq\ & M \int_U p(t)\eta[\lambda(f(t+\cdot) - f(\cdot))]\, d\mu(t) \\ & + M \int_{\Omega\setminus U} p(t)[\eta(2\lambda f(t+\cdot)) + \eta(2\lambda f(\cdot))]\, d\mu(t). \end{aligned}$$

Since $\eta[\lambda(f(t+\cdot) - f(\cdot))] \leq \omega_\eta(\lambda f, U)$ for $t \in U$, we obtain

$$J_1 \leq M\omega_\eta(\lambda f, U) + M \int_{\Omega\setminus U} p(t)[\eta(2\lambda f(t+\cdot)) + \eta(2\lambda f(\cdot))]\, d\mu(t).$$

By subboundedness of η we thus get

$$\begin{aligned} J_1 &\leq M\omega_\eta(\lambda f, U) + M \int_{\Omega\setminus U} p(t)[\eta(2\lambda C f) + \|\ell\|_\infty + \eta(2\lambda f)]\, d\mu(t) \\ &\leq M\omega_\eta(\lambda f, U) + M[2\eta(2\lambda C f) + \|\ell\|_\infty] \int_{\Omega\setminus U} p(t)\, d\mu(t). \end{aligned}$$

The last inequality together with the inequality (3.6) implies (3.4). □

We have now to estimate the remainder term R in (3.4), given by (3.5). We put

$$\begin{aligned} r^{(0)} &= \sup_{u\neq 0} \left| \frac{1}{u} \int_\Omega K(t,u)\, d\mu(t) - 1 \right|, \\ r^{(k)} &= \sup_{1/k \leq |u| \leq k} \left| \frac{1}{u} \int_\Omega K(t,u)\, d\mu(t) - 1 \right|, \end{aligned}$$

for $k = 1, 2, \ldots$ Moreover, we denote for any function $f \in L^0(\Omega)$

$$A_k = \{t \in \Omega : |f(t)| > k\},\ B_k = \{t \in \Omega : |f(t)| < 1/k\},\ C_k = \Omega \setminus (A_k \cup B_k)$$

for $k = 1, 2, \ldots$. There holds the following

Lemma 3.1. *Let ρ be a monotone, J-quasiconvex with a constant $M \geq 1$ modular on $L^0(\Omega)$ and let η be an arbitrary modular on $L^0(\Omega)$. Let K be an $(L,\psi)_0$-Lipschitz Carathéodory kernel function such that $\{\rho, \psi, \eta\}$ is a properly directed triple and let the operator T be defined by* (3.1). *Finally, let $0 < \lambda < 1$ and α be such that $0 < 16\alpha M\|L\|_1 \leq C_\lambda$. Then for an arbitrary set $S \in \Sigma$ and function $f \in L^0_{\rho+\eta}(\Omega)$ there holds*

(a) $R \leq \rho(2\alpha r^{(0)} f)$,

(b) *for* $k = 1, 2, \ldots,$

$$
\begin{aligned}
R \leq\ & M[\eta(\lambda f \chi_{\Omega\setminus S}) + \eta(\lambda f \chi_{S\cap A_k}) + \eta(\lambda f \chi_{S\cap B_k})] \\
& + [\rho(16\alpha f \chi_{\Omega\setminus S}) + \rho(16\alpha f \chi_{S\cap A_k}) + \rho(16\alpha f \chi_{S\cap B_k})] \\
& + \rho(8\alpha r^{(k)} f),
\end{aligned}
$$

where R *is given by (3.5).*

Proof. (a) immediately follows from the inequality $R \leq r^{(0)}|f(s)|$. In order to prove (b), we apply the obvious inequality $\rho(\sum_{j=1}^{n} f_j) \leq \sum_{j=1}^{n} \rho(nf_j)$ with $n = 4$, obtaining

$$
\begin{aligned}
R \leq\ & \rho\left[8\alpha\left|\int_\Omega K(t, f(\cdot)\chi_{\Omega\setminus S}(\cdot))\, d\mu(t) - f(\cdot)\chi_{\Omega\setminus S}(\cdot)\right|\right] \\
& + \rho\left[8\alpha\left|\int_\Omega K(t, f(\cdot)\chi_{S\cap A_k}(\cdot))\, d\mu(t) - f(\cdot)\chi_{S\cap A_k}(\cdot)\right|\right] \\
& + \rho\left[8\alpha\left|\int_\Omega K(t, f(\cdot)\chi_{S\cap B_k}(\cdot))\, d\mu(t) - f(\cdot)\chi_{S\cap B_k}(\cdot)\right|\right] \\
& + \rho\left[8\alpha\left|\int_\Omega K(t, f(\cdot)\chi_{S\cap C_k}(\cdot))\, d\mu(t) - f(\cdot)\chi_{S\cap C_k}(\cdot)\right|\right].
\end{aligned}
$$

Let $P \in \Sigma$ be arbitrary. Applying the assumptions that K is $(L, \psi)_0$-Lipschitz, ρ is monotone, J-quasiconvex and the triple $\{\rho, \psi, \eta\}$ is properly directed, we obtain

$$
\begin{aligned}
& \rho\left[8\alpha\left|\int_\Omega K(t, f(\cdot)\chi_P(\cdot))\, d\mu(t) - f(\cdot)\chi_P(\cdot)\right|\right] \\
& \quad \leq \rho\left[16\alpha \int_\Omega |K(t, f(\cdot)\chi_P(\cdot))|\, d\mu(t)\right] + \rho(16\alpha f \chi_P) \\
& \quad \leq \rho\left[\int_\Omega p(t) 16\alpha \|L\|_1 \psi(t, |f(\cdot)|\chi_P(\cdot))\, d\mu(t)\right] + \rho(16\alpha f \chi_P) \\
& \quad \leq M \int_\Omega p(t)\rho[16\alpha M \|L\|_1 \psi(t, |f(\cdot)|\chi_P(\cdot))\, d\mu(t) + \rho(16\alpha f \chi_P) \\
& \quad \leq M\eta(\lambda f \chi_P) + \rho(16\alpha f \chi_P).
\end{aligned}
$$

Applying the last inequality for $P = \Omega\setminus S$, $P = S\cap A_k$, $P = S\cap B_k$ and the definition of $r^{(k)}$, we obtain the inequality (b). □

3.2 Convergence theorems

Theorem 3.2 and Lemma 3.1 will be the basis for an approximation theorem. We shall approximate functions from a modular space $L^0_{\rho+\eta}(\Omega)$ by means of families of integral operators T_w of the form

$$(T_w f)(s) = \int_\Omega K_w(t, f(t+s))\, d\mu(t), \tag{3.7}$$

where K_w are kernel functions and $w \in W$. Here, W will be an infinite set of indices. For the sake of simplicity we shall limit ourselves to the case when W is an infinite subset of the interval $[a, w_0[$, where $a \in \mathbb{R}$ and $w_0 \in \mathbb{R}$, $w_0 > a$, or $w_0 = +\infty$, and w_0 is a point of accumulation of the set W. Convergence $a_w \overset{w}{\to} a_0$, where $a_w, a_0 \in \mathbb{R}$, $w \in W$, will mean that for every $\varepsilon > 0$ there exists a left neighbourhood U_ε of w_0, equal to an interval $]w_0 - \delta, w_0[$ when $w_0 < +\infty$ and equal to a halfline $[w_1, +\infty[$ in the case when $w_0 = +\infty$, such that $|a_w - a_0| < \varepsilon$ for all $w \in U_\varepsilon \cap W$. In case when $W = [a, w_0[$ and $w_0 < +\infty$ we get $a_0 = \lim_{w \to w_0^-} a_w$, taking $w_0 = +\infty$ we obtain $a_0 = \lim_{w \to +\infty} a_w$ and taking $W = I\!N$ = the set of positive integers, $w_0 = +\infty$, we have $a_0 = \lim_{n \to +\infty} a_n$. Let us remark that a necessary and sufficient condition in order that $a_w \overset{w}{\to} a_0$ is that for any sequence (w_n) such that $w_n \in W$ for $n = 1, 2, \ldots$ and $w_n \to w_0^-$ as $n \to +\infty$ there holds $a_{w_n} \to a_0$ as $n \to +\infty$. We may limit ourselves here to increasing sequences (w_n). At the end let us still remark that most considerations below remain valid in the more general case when W is an abstract set and convergence is meant in the sense of a filter of subsets of W. For some purposes it is also needed that the convergence in the sense of the filter be countably generated.

A family of Carathéodory kernel functions $\mathbb{K} = (K_w)_{w \in W}$, is called a *Carathéodory kernel.* Let $\mathbb{L} = (L_w)_{w \in W}$ be a family of nonnegative functions $L_w \in L^1(\Omega)$ for $w \in W$. We say that the kernel $\mathbb{K} = (K_w)_{w \in W}$ is $(\mathbb{L}, \psi)$*-Lipschitz*, briefly $\mathbb{K} \in (\mathbb{L}, \psi)$-Lip, if $K_w \in (L_w, \psi)$-Lip for $w \in W$ and $D = \sup_{w \in W} \|L_w\|_1 < +\infty$.

Let $\mathcal{U}$ be a filter of subsets of Ω with a basis $\mathcal{U}_0 \subset \Sigma$ (see (2.2)). We write

$$r_w^{(0)} = \sup_{u \neq 0} \left| \frac{1}{u} \int_\Omega K_w(t, u)\, d\mu(t) - 1 \right|,$$

$$r_w^{(k)} = \sup_{1/k \leq |u| \leq k} \left| \frac{1}{u} \int_\Omega K_w(t, u)\, d\mu(t) - 1 \right|$$

for $w \in W$ and $k = 1, 2, \ldots$ We say that the kernel $\mathbb{K}$ is *singular*, if

$$\int_{\Omega \setminus U} p_w(t)\, d\mu(t) \overset{w}{\to} 0 \quad \text{for every } U \in \mathcal{U}_0, \tag{3.8}$$

$\Gamma_k = \sup_{w \in W} r_w^{(k)} < +\infty$ and $r_w^{(k)} \overset{w}{\to} 0$ for $k = 1, 2, \ldots$ If there hold (3.8), $\Gamma_0 = \sup_{w \in W} r_w^{(0)} < +\infty$ and $r_w^{(0)} \overset{w}{\to} 0$, then the kernel $\mathbb{K}$ is called *strongly singular*. Obviously, a strongly singular kernel is singular.

Example 3.2. Let $\mathbb{K} = (K_w)_{w \in W}$ consist of linear kernels, i.e. $K_w(t, u) = \widetilde{K}_w(t) u$. Then

$$\frac{1}{u} \int_\Omega K_w(t, u)\, d\mu(t) = \frac{1}{u} \int_\Omega \widetilde{K}_w(t) u\, d\mu(t) = \int_\Omega \widetilde{K}(t)\, d\mu(t)$$

for $u \neq 0$ and $w \in W$. Thus, for an arbitrary $k = 0, 1, 2, \ldots$ the condition $r_w^{(k)} \xrightarrow{w} 0$ means that

$$\int_\Omega \widetilde{K}_w(t)\, d\mu(t) \xrightarrow{w} 1.$$

This shows that in our case conditions (3.8) and $r_w^{(k)} \xrightarrow{w} 0$ are analogous to those mentioned for Fejér kernel functions K_n in Section 3.1.

Let $\mathbb{T} = (T_w)_{w \in W}$ be the family of operators defined by (3.7). The *domain of* $\mathbb{T}$ is defined as $\operatorname{Dom} \mathbb{T} = \bigcap_{w \in W} \operatorname{Dom} T_w$. We shall prove now the main theorem concerning approximation of functions $f \in L^0_{\rho+\eta}(\Omega) \cap \operatorname{Dom} \mathbb{T}$ by operators of the form (3.7).

Theorem 3.3. *Let $\{\Omega, \mathcal{U}, \Sigma, \mu\}$ be a correctly filtered system and let $+ : \Omega \times \Omega \to \Omega$ be a commutative operation in Ω. Let ρ be a monotone, J-quasiconvex with a constant $M \geq 1$ modular on $L^0(\Omega)$ and let η be a monotone, bounded with respect to the operation $+$, absolutely finite and absolutely continuous modular on $L^0(\Omega)$. Let $\mathbb{K}$ be a singular Carathéodory $(\mathbb{L}, \psi)$-Lipschitz kernel such that $\{\rho, \psi, \eta\}$ is a properly directed triple and let us suppose that one of the following two conditions holds:*

(1) $\mathbb{K}$ *is strongly singular,*

(2) ρ *is finite and absolutely continuous.*

Then for every $f \in L^0_{\rho+\eta}(\Omega) \cap \operatorname{Dom} \mathbb{T}$ there exists an $\alpha > 0$ such that

$$\rho[\alpha(T_w f - f)] \xrightarrow{w} 0.$$

Proof. Let $U \in \Sigma$, $0 < \lambda < 1$ and $0 < \alpha < C_\lambda (2MD)^{-1}$, where $D = \sup_{w \in W} \|L_w\|_1$. Let us write

$$R_w = \rho\left[2\alpha \left| \int_\Omega K_w(t, f(\cdot))\, d\mu(t) - f(\cdot) \right|\right] \quad \text{for } w \in W.$$

By Theorem 3.2, we have

$$\rho[\alpha(T_w f - f)] \leq M\omega_\eta(\lambda f, U) + M[2\eta(2\lambda C f) + \|\ell\|_\infty] \int_{\Omega \setminus U} p_w(t)\, d\mu(t) + R_w$$

for $w \in W$. Let ε be an arbitrary positive number. Since η is a monotone, bounded, absolutely finite and absolutely continuous modular on $L^0(\Omega)$, $\{\Omega, \mathcal{U}, \Sigma, \mu\}$ is a correctly filtered system and $f \in L^0_\eta(\Omega)$, by Theorem 2.4 there holds $\omega_\eta(\lambda f, U) \xrightarrow{\mathcal{U}} 0$ for sufficiently small $\lambda > 0$, and we may assume that $\lambda < 1$. Moreover, we may choose

λ so small that $\eta(2\lambda Cf) < +\infty$, because $f \in L^0_\eta(\Omega)$. Let us fix a set $U \in \mathcal{U}_0$ such that $M\omega_\eta(\lambda f, U) < \varepsilon/4$. By (3.8), we have

$$\int_{\Omega\setminus U} p_w(t)\,d\mu(t) \stackrel{w}{\to} 0,$$

hence there exists a $w_1 \in W$ such that

$$M[2\eta(2\lambda Cf) + \|\ell\|_\infty] \int_{\Omega\setminus U} p_w(t)\,d\mu(t) < \varepsilon/4$$

for $w \in [w_1, w_0[$. Consequently, we obtain

$$\rho[\alpha(T_w f - f)] < \frac{\varepsilon}{2} + R_w \tag{3.9}$$

for sufficiently small $\lambda > 0$ and $\alpha > 0$ and $w \in [w_1, w_0[$. What remains is to show that $R_w \stackrel{w}{\to} 0$.

First, we prove this under the assumption (1). By Lemma 3.1 (a), there holds $R_w \le \rho(2\alpha r_w^{(0)} f)$ for $w \in W$. By the assumption of strong singularity of $\mathbb{K}$, there holds $r_w^{(0)} \stackrel{w}{\to} 0$. Since $f \in L^0_\rho(\Omega)$, we have $\rho(\delta f) \to 0$ as $\delta \to 0^+$, whence there is a $\delta_0 > 0$ such that if $0 < \delta \le \delta_0$, then $\rho(\delta f) < \varepsilon/2$. Now, we take a $w_2 \in W$, $w_2 > w_1$, such that $2\alpha r_w^{(0)} \le \delta_0$ for $w \in [w_2, w_0[$. Then $R_w \le \rho(2\alpha r_w^{(0)} f) < \varepsilon/2$ for $w \in [w_2, w_0[$, $w \in W$. Consequently, $R_w \stackrel{w}{\to} 0$ and $\rho[\alpha(T_w f - f)] < \varepsilon$ for $w \in [w_2, w_0[$, $w \in W$. Thus $\rho[\alpha(T_w f - f)] \stackrel{w}{\to} 0$.

The situation in the case of the assumption (2) is a little more complicated. Again, by (3.9) we have to show that $R_w \stackrel{w}{\to} 0$, applying the estimation (b) from Lemma 3.1 with R_w in place of R. Let $S \in \Sigma$ be an arbitrary subset of finite measure of Ω. Since $A_1 \supset A_2 \supset \cdots$, we have $S \cap A_1 \supset S \cap A_2 \supset \ldots$ and $\mu(S \cap A_1) < +\infty$. Hence $\lim_{k\to+\infty} \mu(S \cap A_k) = \mu(S \cap \bigcap_{k=1}^\infty A_k)$. But by $f \in L^0(\Omega)$, there is a set $\Omega_0 \subset \Omega$, $\Omega_0 \in \Sigma$ of measure $\mu(\Omega_0) = 0$ such that $|f(t)| < +\infty$ for $t \in \Omega \setminus \Omega_0$. From the inclusion $\bigcap_{k=1}^\infty (S \cap A_k) \subset \Omega_0$ we deduce that $\lim_{k\to+\infty} \mu(S \cap A_k) = 0$. Now, applying absolute continuity of both η and ρ we may choose a set S of finite measure and constants $\lambda > 0$ and $\alpha > 0$ so small that

$$M\eta(\lambda f \chi_{\Omega\setminus S}) + \rho(16\alpha f \chi_{\Omega\setminus S}) < \varepsilon/12. \tag{3.10}$$

Keeping S fixed, we may find an index k_0 such that

$$M\eta(\lambda f \chi_{S\cap A_k}) + \rho(16\alpha f \chi_{S\cap A_k}) < \varepsilon/12 \tag{3.11}$$

for all $k \ge k_0$. We also have

$$M\eta(\lambda f \chi_{S\cap B_k}) + \rho(16\alpha f \chi_{S\cap B_k}) \le M\eta\Big(\frac{\lambda}{k}\chi_S\Big) + \rho\Big(\frac{16\alpha}{k}\chi_S\Big).$$

Since η and ρ are finite, the right-hand side of the last inequality tends to 0 as $k \to +\infty$. Hence one may fix an index $k \geq k_0$ such that

$$\eta\left(\frac{\lambda}{k}\chi_S\right) < \varepsilon/24 \quad \text{and} \quad \rho\left(\frac{16\alpha}{k}\chi_S\right) < \varepsilon/24.$$

Consequently, we have for this fixed index k the inequality

$$M\eta(\lambda f \chi_{S \cap B_k}) + \rho(16\alpha f \chi_{S \cap B_k}) < \varepsilon/12.$$

Inserting the last estimation and (3.10) and (3.11) in the inequality (b) from Lemma 3.1, we obtain the inequality

$$R_w \leq \frac{\varepsilon}{4} + \rho(8\alpha r_w^{(k)} f) \tag{3.12}$$

for $w \in W$ and for the above fixed index k. Repeating the argument used in the case (1) to the second term of the above inequality we easily obtain the relation

$$\rho(8\alpha r_w^{(k)} f) \xrightarrow{w} 0.$$

Hence there exists a $w_3 \in W$, $w_3 > w_2$, $w_3 \in [w_2, w_0[$ such that

$$\rho(8\alpha r_w^{(k)} f) < \frac{\varepsilon}{4}$$

for $w \in [w_3, w_0[$, $w \in W$. By the inequality (3.12), we have $R_w < \varepsilon/2$ for such w. Thus, applying the inequalities (3.9) and (3.12), we obtain

$$\rho(\alpha(T_w f - f)) < \varepsilon$$

for $w \in [w_3, w_0[$, $w \in W$. Consequently

$$\rho(\alpha(T_w f - f)) \xrightarrow{w} 0. \qquad \square$$

3.3 Examples

The second part of the assumption of singularity of a kernel $\mathbb{K}$, stating that $r_w^{(k)} \xrightarrow{w} 0$, may be interpreted as a statement that $K_w(t, u)$ behaves nearly as u for w sufficiently near to w_0. Following this way of argument one could try to conclude that if K_w satisfy a generalized Lipschitz condition with a function $\psi(t, u)$, then ψ as a function of u should also behave nearly as u, i.e. the generalized Lipschitz condition is nearly a usual Lipschitz condition with power 1. The next example shows that this kind of argumentation is wrong.

Example 3.3. Let $W = \mathbb{N} =$ the set of positive integers, $w_0 = +\infty$, and let (Ω, Σ, μ) be a measure space and let $\mathcal{U}$ be a filter of subsets of Ω, with a basis $\mathcal{U}_0 \subset \Sigma$. Let $0 < L_n \in L^1(\Omega)$ for $n = 1, 2, \dots$ be such that $\|L_n\|_1 \to 1$ and $\int_{\Omega \setminus U} L_n(t)\, d\mu(t) \to 0$ as $n \to +\infty$ for every $U \in \mathcal{U}_0$ (for example, as L_n we may take the Fejér kernel functions, with $\Omega = [-\pi, \pi]$ endowed with the Lebesgue measure). Let $\mathbb{L} = (L_n)_{n=1}^{\infty}$ and let $H_n : \mathbb{R}_0^+ \to \mathbb{R}_0^+$ be defined by

$$H_n(u) = \left\{ \frac{1}{n} \left(u - \frac{k}{n} \right) \right\}^{1/2} + \frac{k}{n} \quad \text{for } u \in \left[\frac{k}{n}, \frac{k+1}{n} \right[, \ k = 0, 1, 2, \dots$$

and let us extend $H_n(u)$ to the whole $\mathbb{R}$ on putting $H_n(u) = -H_n(-u)$ for $u < 0$, where $n = 1, 2, \dots$ Let $K_n(t, u) = L_n(t) H_n(u)$ for $t \in \Omega$, $u \in \mathbb{R}$, $n = 1, 2, \dots$. It is easily seen that $\mathbb{K} = (K_n)_{n=1}^{\infty}$ is an $(\mathbb{L}, \psi)$- Lipschitz Carathéodory kernel with $\psi(t, u) = \sqrt{u}$ for $t \in \Omega$, $u \geq 0$, but does not satisfy a Lipschitz condition with $\psi(t, u) = u$. Moreover, $\int_{\Omega} K_n(t, u)\, d\mu(t)$ converges as $n \to +\infty$ to u uniformly on every interval $[a, b[\subset \mathbb{R}^+$, where $0 < a < b \in \widetilde{\mathbb{R}}$. Hence one may deduce easily that $\mathbb{K}$ is singular.

Example 3.4. In a similar manner as in Example 3.3 one can also define a strongly singular kernel. For example we modify the definition of $H_n(u)$ from Example 3.3 near the point $u = 0$ in such a way that

$$\left| \frac{H_n(u)}{u} - 1 \right| \leq \frac{1}{n} \quad \text{for } 0 < u \leq \frac{1}{n}.$$

Now we conclude the section with further examples.

Example 3.5. Let $\Omega = [0, 1] \subset \mathbb{R}$ with Lebesgue measure and let us interpret the operation $+$ as usual multiplication. Let $w_r : [0, 1] \to \mathbb{R}_0^+$ for $r \in \mathbb{R}^+ =]0, +\infty[$, with $r_0 = +\infty$. Let us suppose that

$$\int_0^1 w_r(t)\, dt = 1 \quad \text{for every } r \in \mathbb{R}^+,$$
$$\int_0^{1-\delta} w_r(t)\, dt \to 0 \quad \text{as } r \to +\infty$$

for every $\delta \in]0, 1/2[$. We take (see Example 2.5 (b)) $\mathcal{U}_0 = \{[1 - \delta, 1] : \delta \in]0, 1/2[\}$ as a basis of a filter $\mathcal{U}$. Operators T_r defined by

$$(T_r f)(s) = \int_0^1 w_r(t) f(ts)\, dt, \quad r \in \mathbb{R}^+$$

are called average operators or moment operators.

Example 3.6. Let $\Omega = [0, +\infty[= \mathbb{R}_0^+$ be equipped with the operation of usual addition $+$ and let Σ be the σ-algebra of all Lebesgue measurable subsets of Ω and let $g : \Omega \to \mathbb{R}_0^+$ be a locally integrable function. We put $\mu(A) = \int_A g(t)\,dt$ for an arbitrary set $A \in \Sigma$. Obviously, μ is a σ-finite measure on Σ. Let $\mathcal{U}$ be the filter of all right-neighbourhoods of zero in Ω. A basis $\mathcal{U}_0$ of $\mathcal{U}$ consisting of Σ-measurable sets may be defined e.g. by $\mathcal{U}_0 = \{[0, 1/n) : n = 1, 2, \dots\}$. Then $(\Omega, \mathcal{U}, \Sigma, \mu)$ is a correctly filtered system, and a family of convolution-type operators $(T_w)_{w\in W}$ is defined by

$$(T_w f)(s) = \int_0^{+\infty} K_w(t, f(t+s))\,d\mu(t).$$

Example 3.7. Let $\Omega = \mathbb{N}_0 = \{0, 1, 2, \dots\}$, $\Sigma = 2^\Omega$ and let μ be the counting measure on $\mathbb{N}_0$. Now, the function f is a sequence $(t_i)_{i=0}^\infty$. Let $\rho(f) = \sum_{i=0}^\infty |t_i|$ for any f and let η be a modular on Ω. As $\mathcal{U}$ we can take the family of complements of finite (or empty) sets (see Example 2.5 (c)). For $w \in W$ we may define operators $T_w f = ((T_w f)_j)_{j=0}^\infty$, of the form

$$(T_w f)_j = \sum_{i=0}^\infty K_{w,i}(t_{i+j}),$$

with $K_{w,i} : \mathbb{R} \to \mathbb{R}$. Singularity of the kernel $\mathbb{K}$ defined by the kernel functions $K_w : \mathbb{N}_0 \times \mathbb{R} \to \mathbb{R}$ means that

$$\sup_{1/k \le |u| \le k} \left| \frac{1}{u} \sum_{i=0}^\infty K_{w,i}(u) - 1 \right| \xrightarrow{w} 0,$$

for $k = 1, 2, \dots$, and $p_{w,i} \xrightarrow{w} 0$ for every $i = 0, 1, 2 \dots$, where

$$p_{w,i} = L_{w,i} \Big[\sum_{i=1}^\infty L_{w,i} \Big]^{-1}$$

are the normalized functions L_w given in the Lipschitz condition of K_w.

3.4 Rates of modular approximation in modular Lipschitz classes

Let $\{\Omega, \Sigma, \mu\}$ be a (Hausdorff topological) measure space with a σ-finite, complete measure, $+ : \Omega \times \Omega \to \Omega$ is a commutative operation with unity θ. Let $\mathcal{U}$ be the filter of neighbourhoods of θ. As before we will suppose that $\mathcal{U}$ contains a base $\mathcal{U}_0$ of measurable subsets of Ω.

Let ρ, η be two modulars on $L^0(\Omega)$ and let $L^0_\rho(\Omega)$, $L^0_\eta(\Omega)$ be the modular spaces generated by ρ and η, respectively.

Let $\mathcal{T}$ be the class of measurable functions $\tau : \Omega \to \mathbb{R}_0^+$ such that $\tau(t) > 0, t \in \Omega$, $t \neq \theta$. For a given $\tau \in \mathcal{T}$, we define the subspace $\mathrm{Lip}_\tau(\rho)$ of L^0_ρ by

$$\mathrm{Lip}_\tau(\rho) = \{f \in L^0_\rho(\Omega) : \text{there is } \alpha > 0 \text{ with } \rho[\alpha|f(\cdot+t) - f(\cdot)|] = \mathcal{O}(\tau(t)), t \to \theta\},$$

where, for any two functions $f, g \in L^0(\Omega)$, $f(t) = \mathcal{O}(g(t)), t \to \theta$, means that there is a constant $B > 0$ and $U \in \mathcal{U}_0$ such that $|f(t)| \leq B|g(t)|$ for $t \in U$.

Analogously we define the class $\mathrm{Lip}_\tau(\eta)$. We call these classes *modular Lipschitz classes*.

Example 3.8. Let $\rho(f) = \|f\|_p$, the L^p-norm of a function $f \in L^p(\Omega)$, $p \geq 1$. Then we have

$$\mathrm{Lip}_\tau(\rho) = \{f \in L^0_\rho(\Omega) : \|f(\cdot + t) - f(\cdot)\|_p = \mathcal{O}(\tau(t)), t \to \theta\}.$$

In particular if $\Omega = \mathbb{R}$ with Lebesgue measure and $\tau(t) = t^\alpha$, with $\alpha > 0$, we obtain the classical Zygmund classes for L^p-spaces.

We will obtain some results about the rates of modular approximation of a family of nonlinear integral operators, whose kernels satisfy an (L, ψ)- Lipschitz condition.

Let $\psi \in \Psi$ be a given function (see Section 3.1) and let $W \subset \mathbb{R}$ be a set of indices, as in Section 3.2, and $\{K_w\}_{w \in W}$ a $(\mathbb{L}, \psi)$-Lipschitz Carathéodory kernel. Let Ξ be the class of all functions $\xi : W \to \mathbb{R}^+$ such that $\xi(w) \to 0$ as $w \to w_0^-$.

For a given $\xi \in \Xi$, we will say that $(K_w)_{w \in W}$ is *strongly ξ-singular* if

(i) for every $U \in \mathcal{U}_0$, we have

$$\int_{\Omega \setminus U} p_w(t)\, d\mu(t) = \mathcal{O}(\xi(w)), \quad w \to w_0^-,$$

(ii) $\Gamma_0 = \sup_{w \in W} r_w^{(0)} < +\infty$, and $r_w^{(0)} = \mathcal{O}(\xi(w)), \quad w \to w_0^-$.

For the given kernel $(K_w)_{w \in W}$ we denote by $\mathbb{T} = (T_w)_{w \in W}$ the family of integral operators

$$(T_w f)(s) = \int_\Omega K_w(t, f(s+t))\, d\mu(t), \quad s \in \Omega,$$

for $f \in \mathrm{Dom}\,\mathbb{T}$.

We have the following

Theorem 3.4. *Let $\tau \in \mathcal{T}$ and $\xi \in \Xi$ be fixed. Let ρ be a monotone, quasiconvex and J-quasiconvex with a constant $M \geq 1$ modular on $L^0(\Omega)$, and η be a monotone and subbounded modular on $L^0(\Omega)$, such that the triple $\{\rho, \psi, \eta\}$ is properly directed.*

Let $\{K_w\}$ *be a* ξ*-strongly singular,* $(\mathbb{L}, \psi)$*-Lipschitz Carathéodory kernel and* $f \in \mathrm{Lip}_\tau(\eta) \cap L^0_\rho$. *Let us suppose that there is* $U \in \mathcal{U}_0$ *such that*

$$\int_U p_w(t)\tau(t)\,d\mu(t) = \mathcal{O}(\xi(w)), \quad w \to w_0^-. \tag{3.13}$$

Then for sufficiently small $a > 0$,

$$\rho[a(T_w f - f)] = \mathcal{O}(\xi(w)), \quad w \to w_0^-.$$

Proof. For $f \in \mathrm{Lip}_\tau(\eta)$ let $\alpha > 0$, $U \in \mathcal{U}_0$, and $R \geq 0$, be such that

$$\eta[\alpha|f(t + \cdot) - f(\cdot)|] \leq R\tau(t),$$

for every $t \in U$. Let $R' \geq 0$ be such that

$$r_w^{(0)} \leq R'\xi(w),$$

for $w \in W$, sufficiently near w_0. Let $\lambda > 0$ be such that $\eta(2\lambda C f) < +\infty$, where C is the constant from the subboundedness of η, and $\lambda < \alpha$. Let $a > 0$ be such that $2aDM \leq C_\lambda$, where $D = \sup_{w \in W} \|L_w\|_1$, M is the constant of the J-quasiconvexity of ρ and C_λ is the constant of the definition of properly directed triple, and such that $\rho[2aMR'f] < +\infty$. We have

$$\begin{aligned}\rho[a(T_w f - f)] \leq \rho\left[2a \int_\Omega |K_w(t, f(t + \cdot)) - K_w(t, f(\cdot))|\,d\mu(t)\right] \\ + \rho\left[2a \left|\int_\Omega K_w(t, f(\cdot))\,d\mu(t) - f(\cdot)\right|\right] = J_1 + J_2.\end{aligned}$$

We now evaluate J_1. From the J-quasiconvexity of ρ, the choice of $a > 0$ and the assumption on the triple $\{\rho, \psi, \eta\}$, we have

$$\begin{aligned}J_1 &\leq M \int_\Omega p_w(t)\eta[\lambda|f(t + \cdot) - f(\cdot)|]\,d\mu(t) \\ &= M\left\{\int_U + \int_{\Omega\setminus U}\right\} p_w(t)\eta[\lambda|f(t + \cdot) - f(\cdot)|]\,d\mu(t) = J_1^1 + J_1^2.\end{aligned}$$

If we choose U in such a way that (3.13) holds, and taking into account that $\lambda < \alpha$, we have

$$J_1^1 \leq MR \int_U p_w(t)\tau(t)\,d\mu(t) = \mathcal{O}(\xi(w)), \quad w \to w_0^-,$$

and from the subboundedness of η, we obtain

$$J_1^2 \leq M\{\|\ell\|_\infty + 2\eta[2\lambda C f]\} \int_{\Omega\setminus U} p_w(t)\,d\mu(t) = \mathcal{O}(\xi(w)), \quad w \to w_0^-.$$

Next, we consider J_2. For every w, sufficiently near w_0, we have

$$J_2 \leq \rho[2aR'\xi(w)f].$$

Now, since $\xi(w) \stackrel{w}{\to} 0$, we have that $\xi(w) < 1$ for $w \in W$, sufficiently near w_0. Thus, by quasiconvexity of ρ, we finally deduce

$$J_2 \leq M\xi(w)\rho[2aMR'f] = \mathcal{O}(\xi(w)), \quad w \to w_0^-.$$

The assertion is now proved. □

Example 3.9. Let $\Omega = \mathbb{R}$ with Lebesgue measure. We recall that given a function $L : \mathbb{R} \to \mathbb{R}$, $L \in L^1(\mathbb{R})$, the *α-absolute moment of L* is defined by the integral

$$\nu_\alpha = \int_{\mathbb{R}} |t|^\alpha |L(t)|\, dt,$$

if it is finite. The assumption (3.13) is linked to the existence of moments. Indeed, let $W = [1, +\infty[$, and $w_0 = +\infty$ and let us suppose that the functions $(p_w)_w$ are of type

$$p_w(t) = wL(wt),$$

for every $w \in W$, where $L : \mathbb{R} \to \mathbb{R}_0^+$ is a fixed Lebesgue integrable function, with $\|L\|_1 = 1$. Let us take $\tau(t) = |t|^\alpha$, and $\xi(w) = w^{-\alpha}$, for $\alpha > 0$. Then if $\nu_\alpha < +\infty$, (3.13) holds. Indeed, for $U =]-\delta, \delta[$ we have

$$w^\alpha \int_U wL(wt)|t|^\alpha\, dt = \int_{-\delta w}^{\delta w} L(z)|z|^\alpha\, dz \leq \nu_\alpha < +\infty,$$

that is the assertion.

3.5 Nonlinear perturbations of linear integral operators

In this section we apply the methods developed in this chapter to approximation properties of a family of nonlinear operators given by the sum of a linear integral operator and a nonlinear perturbation, generated by a Carathéodory kernel function (the perturbating kernel).

Let $\{\Omega, \mathcal{U}, \Sigma, \mu\}$ be a correctly filtered system, with respect to a commutative operation $+$, and let $\mathcal{U}_0$ be a basis of the filter $\mathcal{U}$ with $\mathcal{U}_0 \subset \Sigma$.

We will consider integral operators of the form

$$(Tf)(s) = (\widetilde{K} * f)(s) + (Pf)(s), \tag{3.14}$$

where

$$(\widetilde{K} * f)(s) = \int_\Omega \widetilde{K}(t) f(s+t)\, d\mu(t)$$

and

$$(Pf)(s) = \int_\Omega k(t, f(t+s))\, d\mu(t)$$

for $f \in L^0(\Omega)$ belonging to some modular function spaces. We suppose that $\widetilde{K} : \Omega \to \mathbb{R}$, belongs to $L^1(\Omega)$, and $\|\widetilde{K}\|_1 > 0$, and moreover $k : \Omega \times \mathbb{R} \to \mathbb{R}$ is a Carathéodory kernel function (see Section 1.1). We will call $\widetilde{K}$ the *linear kernel function* and k the *perturbating kernel function*. We will assume a $(L, \psi)_0$-Lipschitz condition on k, where $\psi : \Omega \times \mathbb{R}_0^+ \to \mathbb{R}_0^+$ belongs to the class Ψ (see Section 3.1), and $L : \Omega \to \mathbb{R}_0^+$ is a function in $L^1(\Omega)$, with $\|L\|_1 > 0$. As before, we will put $p(t) = L(t)/\|L\|_1$.

We begin with the following embedding theorem.

Theorem 3.5. *Let ρ, η be two J-quasiconvex with the same constant $M \geq 1$ modulars on $L^0(\Omega)$, subbounded with constants C', C'' and functions ℓ, ℓ' respectively. Let the perturbating kernel function k be $(L, \psi)_0$-Lipschitz with $L \in L^1(\Omega)$. Suppose that $\{\rho, \psi, \eta\}$ is a properly directed triple, and the operator T be defined by* (3.14). *Let $0 < \lambda < 1$ and let $0 < \alpha \leq (1/2)\min(\lambda/(M\|\widetilde{K}\|_1), C_\lambda/(M\|L\|_1))$. Then for every function $f \in L^0_{\rho+\eta}(\Omega) \cap \operatorname{Dom} T$ there holds the inequality*

$$\rho(\alpha T f) \leq M[\rho(\lambda C' f) + \eta(\lambda C'' f) + \|\ell\|_\infty + \|\ell'\|_\infty].$$

Proof. Since ρ is monotone, J-quasiconvex with constant $M \geq 1$, and subbounded, and $\alpha > 0$ is such that $2\alpha M\|\widetilde{K}\|_1 \leq \lambda$ with $0 < \lambda < 1$, we obtain

$$\begin{aligned}\rho(2\alpha(\widetilde{K} * f)) &\leq \frac{M}{\|\widetilde{K}\|_1}\int_\Omega \widetilde{K}(t)\rho(\lambda|f(t+\cdot)|)\, d\mu(t)\\ &\leq M[\rho(\lambda C' f) + \|\ell\|_\infty].\end{aligned}$$

Since k is $(L, \psi)_0$-Lipschitz, the triple $\{\rho, \psi, \eta\}$ is properly directed, and α is taken such that $2\alpha M\|L\|_1 \leq C_\lambda$, we get, again by J-quasiconvexity of ρ with constant $M \geq 1$ and by subboundedness of η,

$$\begin{aligned}\rho(2\alpha(Pf)) &\leq \frac{M}{\|L\|_1}\int_\Omega L(t)\eta(\lambda f(t+\cdot))\, d\mu(t)\\ &\leq M[\eta(\lambda C'' f) + \|\ell'\|_\infty].\end{aligned}$$

We get finally

$$\begin{aligned}\rho(\alpha T f) &\leq \rho(2\alpha((\widetilde{K} * f)) + \rho(2\alpha(Pf))\\ &\leq M[\rho(\lambda C' f) + \eta(\lambda C'' f) + \|\ell\|_\infty + \|\ell'\|_\infty].\end{aligned}$$ □

As a corollary we may conclude immediately

Corollary 3.2. *Under the assumptions of Theorem 3.5, T maps $L^0_{\rho+\eta}(\Omega) \cap \operatorname{Dom} T$ into $L^0_\rho(\Omega)$. If moreover ρ and η are strongly subbounded, if $f_n \in L^0_{\rho+\eta}(\Omega) \cap \operatorname{Dom} T$ and $f_n \xrightarrow{\rho+\eta} 0$, then $Tf_n \xrightarrow{\rho} 0$.*

Next we will estimate the error of modular approximation of f by Tf (compare with Theorem 3.2).

Theorem 3.6. *Let ρ be a monotone modular on $L^0(\Omega)$, J-quasiconvex with constant $M \geq 1$, subbounded with constant C' and function ℓ and let η be a subbounded modular on $L^0(\Omega)$ with constant C'' and function ℓ'. Let the linear kernel function $\widetilde{K}$ be non-negative and let the perturbating kernel function k be $(L, \psi)_0$-Lipschitz. Moreover, let $\{\rho, \psi, \eta\}$ be a properly directed triple. Let $0 < \lambda < 1$, $2\alpha M \|L\|_1 \leq C_\lambda$, $U \in \mathcal{U}_0$ and $f \in L^0_{\rho+\eta}(\Omega) \cap \operatorname{Dom} T$. Then there holds the inequality*

$$\begin{aligned} \rho(\alpha(Tf - f)) \leq{} & M\omega_\rho(4\alpha M\|\widetilde{K}\|_1 f, U) + 2M[\rho(8\alpha C' M\|\widetilde{K}\|_1 f) \\ & + \frac{1}{2}\|\ell\|_\infty] \int_{\Omega\setminus U} \frac{\widetilde{K}(t)}{\|\widetilde{K}\|_1}\, d\mu(t) + M[\eta(\lambda C'' f) + \|\ell'\chi_U\|_\infty \\ & + \|\ell'\|_\infty \int_{\Omega\setminus U} p(t)\, d\mu(t)] + \rho[4\alpha(\|\widetilde{K}\|_1 - 1)f]. \end{aligned} \tag{3.15}$$

Proof. For every α we have

$$\rho(\alpha(Tf - f)) \leq \rho[2\alpha(\widetilde{K} * f - f)] + \rho[2\alpha(Pf)]. \tag{3.16}$$

Applying the identity

$$(\widetilde{K} * f - f)(s) = \int_\Omega \widetilde{K}(t)(f(t + s) - f(s))\, d\mu(t) + (\|\widetilde{K}\|_1 - 1)f(s),$$

by monotonicity of ρ we obtain

$$\rho[2\alpha(\widetilde{K} * f - f)] \leq J_1 + J_2, \tag{3.17}$$

where

$$J_1 = \rho(4\alpha \int_\Omega \widetilde{K}(t)(f(t + \cdot) - f(\cdot))\, d\mu(t)), \quad J_2 = \rho(4\alpha(\|\widetilde{K}\|_1 - 1)f). \tag{3.18}$$

Now following essentially the same methods used in Theorems 3.2 and 3.5, we have, by J-quasiconvexity with constant $M \geq 1$ and subboundedness of ρ (the details are left to the reader)

$$\begin{aligned} J_1 \leq{} & M\omega_\rho(4\alpha M\|\widetilde{K}\|_1 f, U) \\ & + 2M[\rho(8\alpha C' M\|\widetilde{K}\|_1 f) + \frac{1}{2}\|\ell\|_\infty] \int_{\Omega\setminus U} \frac{\widetilde{K}(t)}{\|\widetilde{K}\|_1}\, d\mu(t). \end{aligned} \tag{3.19}$$

Moreover, since ρ is monotone and J-quasiconvex, the perturbating kernel is $(L, \psi)_0$-Lipschitz, $\{\rho, \psi, \eta\}$ is a properly directed triple and η is subbounded, we obtain for $U \in \mathcal{U}_0$, $0 < \lambda < 1$ and $0 < 2\alpha M\|L\|_1 \leq C_\lambda$

$$\begin{aligned}\rho(2\alpha(Pf)) &\leq M[\eta(\lambda C'' f) + \int_U p(t)\ell'(t)\,d\mu(t) + \int_{\Omega\setminus U} p(t)\ell'(t)\,d\mu(t)] \\ &\leq M\left[\eta(\lambda C'' f) + \|\ell'\chi_U\|_\infty + \|\ell'\|_\infty \int_{\Omega\setminus U} p(t)\,d\mu(t)\right].\end{aligned}$$

Taking together the last inequality and inequalities (3.16), (3.17), (3.18) and (3.19) we obtain (3.15). □

Finally, we take a family $\mathbb{T} = (T_w)_{w\in W}$ of operators T_w of the form (3.14), where W is a subset of $\mathbb{R}$, as in Section 3.2, and let w_0 a point of accumulation of W. Thus we have

$$(T_w f)(s) = (\widetilde{K}_w * f)(s) + (P_w f)(s)$$

for μ- a.e. $s \in \Omega$ (the exceptional set of measure zero is supposed independent of the index w). The family of functions $\widetilde{K}_w : \Omega \to \mathbb{R}_0^+$, $\widetilde{\mathbb{K}} = (\widetilde{K}_w)_{w\in W}$, is called a *linear kernel*, and the family $\mathbb{K} = (k_w)_{w\in W}$ is called a *perturbating kernel*. The domain Dom $\mathbb{T}$ of $\mathbb{T}$ is defined in Section 3.2, i.e. Dom $\mathbb{T} = \bigcap_{w\in W}$ Dom T_w. The definition of an $(\mathbb{L}, \psi)_0$-Lipschitz kernel $\mathbb{K}$ for a family of non-negative functions $L_w \in L^1(\Omega)$ is that introduced in Section 3.2.

We say that the linear kernel $\widetilde{\mathbb{K}}$ is *singular* if the following two conditions hold:

(1) for every $U \in \mathcal{U}_0$ we have

$$\frac{1}{\|\widetilde{K}_w\|_1} \int_{\Omega\setminus U} \widetilde{K}_w(t)\,d\mu(t) \xrightarrow{w} 0,$$

(b) the set $\{\|\widetilde{K}\|_1 : w \in W\}$ is bounded and

$$\|\widetilde{K}_w\|_1 \xrightarrow{w} 1.$$

For the perturbating kernel $\mathbb{K}$ we do not need the singularity, but we simply assume that the family $\{\|L_w\|_1 : w \in W\}$ is bounded and for any $U \in \mathcal{U}_0$ we have

$$\frac{1}{\|L_w\|_1} \int_{\Omega\setminus U} L_w(t)\,d\mu(t) \xrightarrow{w} 0. \tag{3.20}$$

Now we formulate a theorem concerning ρ-convergence $T_w f \xrightarrow{w} f$.

Theorem 3.7. *Let ρ be a J-quasiconvex with a constant $M \geq 1$, monotone, absolutely finite, absolutely continuous, bounded modular on $L^0(\Omega)$. Let η be a bounded modular* *on* $L^0(\Omega)$. *Let* $\widetilde{K} = (\widetilde{K}_w)_{w\in W}$ *be a singular linear kernel, where* $K_w(t) \geq 0$ *for* $t \in \Omega$, *and let* $\mathbb{K} = (k_w)_{w\in W}$ *be an* $(\mathbb{L}, \psi)_0$*-Lipschitz perturbating kernel such that* (3.20) *is satisfied. Let* $\{\rho, \psi, \eta\}$ *be a properly directed triple and let* $f \in L^0_{\rho+\eta}(\Omega) \cap \operatorname{Dom}\mathbb{T}$. *Then there exists a* $\lambda > 0$ *such that*

$$\rho[\lambda(T_w f - f)] \overset{w}{\to} 0.$$

Proof. The proof may be worked with the same arguments used in Theorem 3.3, using now the inequality of Theorem 3.6. The details are left to the reader. □

3.6 Bibliographical notes

Example 3.1 shows the background of the idea of convolution-type nonlinear integral operator, namely some concrete linear integral operators used in classical approximation by trigonometric polynomials (see e.g. [67], compare also [157]). In the general case, linearity of the kernel function $K(t, u)$ with respect to the variable $u \in \mathbb{R}$ is replaced by an (L, ψ)-Lipschitz condition with respect to u, expressed by the inequality (3.2); this condition plays a key role in results presented in this book. In case we are not interested in approximation, but in embedding-type inequalities only, the (L, ψ)-Lipschitz condition may be replaced by the weaker $(L, \psi)_0$-Lipschitz condition, defined by means of the inequality (3.3). In earlier results (see [154], [155]) the (L, ψ)-Lipschitz condition was taken in a strong form, namely with the function $\psi(u) = |u|$, i.e. a usual Lipschitz condition. This was replaced by an arbitrary increasing, concave function ψ with $\psi(0) = 0$ (see [156]). Further progress was obtained in [17], where ψ is supposed to be of the form $\psi(t, u)$, as in the inequalities (3.2) and (3.3) (see also [158]). A study of functions in metric spaces may be found in [172].

Theorem 3.1 concerning some embedding-type inequality was first given in [19] as Proposition 1. Theorem 3.2 and Lemma 3.1 are theorems on modular approximation of f by means of Tf, where Tf is given by the convolution-type operator (3.1) and give estimates of the error of approximation. The first version of these results was obtained already in [155], with estimation of the remainder term R as in Lemma 3.1 (a). The estimation of R of the form as in Lemma 3.1 (b) was obtained first in [19]. Theorem 3.2 was extended to the case of vector-valued functions in [23].

Families of linear integral operators of the form (3.7) were first investigated in [150] in the case when $W = \mathbb{N} =$ the set of positive integers. These investigations were generalized in [151] to the case when W is an arbitrary filtered set of indices. This degree of generality was also maintained in [152] by considering the approximation by families of nonlinear integral operators, and in subsequent papers, as [155], [19] and [22]. Since in applications it suffices to consider the special kind of the set W with its natural convergence, presented in 3.2, we limit ourselves in this book to the case when W is an infinite subset of an interval $[a, w_0[\subset \mathbb{R}$, w_0 being an accumulation point of W, w_0 finite or $+\infty$. There is no problem with generalization of theorems presented here to the case of an abstract filtered set W of indices.

The aim of Section 3.2 is to obtain sufficient conditions in order that the right-hand side of the inequality (3.4), i.e. the error of modular approximation of f by means of $T_w f$ with K replaced by K_w, converges to zero as $w \to w_0$. Analogously as in the case of approximation by linear integral operators, this requires the notions of singularity or strong singularity of the kernel $\mathbb{K} = (K_w)_{w \in W}$ (a family of kernel functions), as defined just before Example 3.2. A singular kernel $\mathbb{K}$ was defined and applied in [19] and a strongly singular kernel in [155]. The fundamental Theorem 3.3 was given in [19] as Theorem 2. Examples 3.3–3.4, which show that the assumption of singularity or strong singularity are not contradictory with (L, ψ)-Lipschitz condition with a function ψ such that $u^{-1}\psi(t, u) \to +\infty$ as $u \to 0^+$, may be found in [22].

The notion of rates of convergence of approximation was developed by P. L. Butzer in various classical papers (see the monograph [67]), with respect to various kinds of convergence (pointwise, uniform or in L^p-norm). A fundamental tool for this theory is given by Lipschitz classes of functions. For pointwise or uniform convergence they are based on the classical Hölder condition of functions, while in L^p-spaces they are defined by means the Zygmund conditions.

The theory of the degree of approximation for linear integral operators, takes its origin from the study of the rapidity of convergence of Fourier series, (see the classical papers by Hardy and Littlewood [115], Bernstein [45]) and in the determination of the saturation classes in the theory of approximation (see e.g. [51], [56]). For a wide bibliography on this subject see [67]. More recent contributions in this direction were given in [47], [180], [181], [9], for uniform or pointwise approximation. Finally, it is important to mention here a series of papers by P. L. Butzer and his school on the use of the Banach–Steinhaus principle in the study of the rate of approximation for linear operators (see e.g. [54], [89], [90]).

Based on the definition of the classical Zygmund class of L^p-functions, the notion of modular Lipschitz class $\mathrm{Lip}_\tau(\rho)$ was introduced in [38], (see Section 3.4), and a result for linear integral operators of convolution type was given.

For families of nonlinear operators first results in this direction were given again in [38] in the case when the function f belongs to modular Lipschitz classes. The results in Section 3.4 were given in [38].

In Section 3.5 we consider the case when the nonlinear integral operator T is a sum of a linear operator defined by means of a linear kernel function $\widetilde{K}$ and a nonlinear operator generated by a perturbating kernel function k. Theorems 3.5, 3.6 and 3.7 are analogous of Theorem 3.1, 3.2, and 3.3, respectively (see [162]).

Chapter 4

Urysohn integral operators with homogeneous kernel functions. Applications to nonlinear Mellin-type convolution operators

4.1 General concepts of homogeneity and Lipschitz conditions

Throughout this chapter, Ω will be a locally compact (and σ-compact) topological group G, provided with its (regular) Haar measure, denoted by μ. Here, for the sake of simplicity, we suppose G abelian. Let θ be the neutral element of G and $\mathcal{U}$ a base of measurable neighbourhoods of $\theta \in G$. Let $\mathcal{X} = L^0(G)$, the space of all the measurable functions $f : G \to \widetilde{\mathbb{R}}$, finite a.e. in G.

Let ρ be a monotone modular on $L^0(G)$, and let $L^0_\rho(G)$ be the corresponding modular space.

Let $\zeta : G \to \mathbb{R}^+$ and $L : G \times G \to \mathbb{R}^+_0$ be measurable functions.

We say that L is *ζ-subhomogeneous* if there is a constant $R \geq 1$ such that

$$\zeta(t)L(s+v, t+v) \leq R\zeta(t+v)L(s,t),$$

for every $s, t, v \in G$.

We denote by $\mathcal{L}_\zeta$ the class of all the ζ-subhomogeneous functions L.

If L is ζ-subhomogeneous with $R = 1$, it is easy to see that we have now

$$\zeta(t)L(s+v, t+v) = \zeta(t+v)L(s,t),$$

for every $s, t, v \in G$. In this case we say that L is *ζ-homogeneous* and we will denote by $\widetilde{\mathcal{L}}_\zeta$ the class of all ζ-homogeneous functions L.

Example 4.1. (a) Let $G = (\mathbb{R}^+, \cdot)$ be the multiplicative group of positive real numbers. If $L : \mathbb{R}^+ \times \mathbb{R}^+ \to \mathbb{R}^+_0$ is homogeneous of degree $\alpha \in \mathbb{R}$, with respect to $(s,t) \in \mathbb{R}^+ \times \mathbb{R}^+$, then L is ζ-homogeneous with $\zeta(t) = t^\alpha$, $t \in \mathbb{R}^+$. In particular we can take the *modified moment kernels* defined, for any positive integers $n = 1, 2, \dots$, by

$$K_n(s,t) = (n+1)\frac{t^{n+1}}{s^{\beta+n+1}}\chi_{]0,s[}(t),$$

for a fixed $\beta \in \mathbb{R}$. This function is homogeneous of degree $-\beta$. For $\beta = 0$, we obtain the classical moment operators.

(b) Let $G = (\mathbb{R}, +)$ be the additive line group. If $L(s,t) = H(s-t)$, with H measurable, then L is homogeneous of degree 0, and this implies that L is ζ-homogeneous with $\zeta(t) = 1$, for every $t \in G$.

(c) Let $G = (\mathbb{R}^N_+, \bullet)$, where $\mathbb{R}^N_+ = (]0, +\infty[)^N$ and for $x = (x_1, \dots, x_N)$, $y = (y_1, \dots, y_N) \in \mathbb{R}^N_+$, we define $x \bullet y = (x_1 y_1, \dots, x_N y_N)$. Thus G is a locally compact topological abelian group, with neutral element $\theta = \mathbf{1} = (1, \dots, 1)$, and for $x = (x_1, \dots, x_N)$ the inverse of x is given by $x^{-1} = (x_1^{-1}, \dots, x_N^{-1})$.

Putting for every $x \in \mathbb{R}^N_+$, $x = (x_1, \dots, x_N)$, $\langle x \rangle = \prod_{k=1}^N x_k$, the Haar measure μ on G is given (up to multiplicative factors), by

$$d\mu(x) = \frac{dx}{\langle x \rangle},$$

where dx is the usual Lebesgue measure.

Let $\alpha = (\alpha_1, \dots, \alpha_N)$ be a fixed multi-index. Denoting $x^\alpha = (x_1^{\alpha_1}, \dots x_N^{\alpha_N})$, for $x = (x_1, \dots, x_N)$, we say that a measurable function $L : G \times G \to \mathbb{R}$ is homogeneous of degree α if

$$L(xv, yv) = \langle v^\alpha \rangle L(x, y),$$

for every $x, y, v \in \mathbb{R}^N_+$. Then putting $\zeta(v) = \langle v^\alpha \rangle$, we obtain a class of ζ-homogeneous functions. A simple example is given by the "box" functions of type

$$L(x, y) = L_1(x_1, y_1) L_2(x_2, y_2) \dots L_N(x_N, y_N),$$

for $x, y \in \mathbb{R}^N_+$, where L_j is α_j-homogeneous with respect (x_j, y_j), $j = 1, \dots, N$.

(d) For $G = (\mathbb{R}^+, \cdot)$ let $\xi : G \to \mathbb{R}^+$ be a measurable function. Let $H : G \times G \to \mathbb{R}^+_0$ be a measurable function homogeneous of degree $\alpha \in \mathbb{R}$. Then the kernel $K(s,t) = \xi(t) H(s,t)$ is ζ-homogeneous with respect to $\zeta(t) = t^\alpha \xi(t)$.

(e) Let $G = (\mathbb{Z}, +)$ be the additive group of integers, and let $F : \mathbb{R}^+ \times \mathbb{R}^+ \to \mathbb{R}$ be a measurable function which is homogeneous of degree $\alpha \in \mathbb{R}$. For a fixed positive constant $T > 0$, let us define the function $L : G \times G \to \mathbb{R}$ given by

$$L(m, n) = F(e^{m/T}, e^{n/T}), \quad m, n \in \mathbb{Z}.$$

Then it is easy to show that L is homogeneous with respect to the function $\zeta : \mathbb{Z} \to \mathbb{R}^+$ defined by $\zeta(k) = \exp(k\alpha/T)$, $k \in \mathbb{Z}$.

Let $K : G \times G \times \mathbb{R} \to \mathbb{R}$ be a measurable function. We will say that K is $(L, 1)$-Lipschitz if there is a non-negative measurable function $L : G \times G \to \mathbb{R}^+_0$ such that

$$|K(s,t,u) - K(s,t,v)| \leq L(s,t)|u - v|,$$

for every $(s,t) \in G \times G$ and $u, v \in \mathbb{R}$.

Let now Ψ' be the class of all the functions $\psi : G \times \mathbb{R}^+_0 \to \mathbb{R}^+_0$ such that the following assumptions hold (compare with Sections 1.4, 3.1):

(1) $\psi(t,\cdot)$ is a continuous, nondecreasing function, for every $t \in G$,

(2) $\psi(t,0) = 0$, $\psi(t,u) > 0$, for $u > 0$ and for every $t \in G$,

(3) ψ is τ*-bounded*, i.e., there are two constants $E_1, E_2 \geq 1$ and a measurable function $F : G \times G \to \mathbb{R}_0^+$ such that

$$\psi(s+z,u) \leq E_1\psi(z,E_2u) + F(s,z),$$

for every $u \in \mathbb{R}_0^+$, $s, z \in G$.

Note that if $\psi(t,u) = \psi(u)$, i.e. ψ is independent of the parameter $t \in G$, then (3) is obviously satisfied with $E_1 = E_2 = 1$ and $F \equiv 0$.

According to the definition given in Section 3.1, we say that K is (L,ψ)-Lipschitz if there is a non- negative measurable function $L : G \times G \to \mathbb{R}_0^+$ such that

$$|K(s,t,u) - K(s,t,v)| \leq L(s,t)\psi(t,|u-v|),$$

for every $(s,t) \in G \times G$ and $u, v \in \mathbb{R}$.

We will denote by $\mathcal{K}_\zeta^1$ the class of all the measurable functions

$$K : G \times G \times \mathbb{R} \to \mathbb{R}$$

such that

(i) $K(s,t,0) = 0$,

(ii) K is an $(L,1)$-Lipschitz function, where L satisfies the following assumptions:

(a) $L \in \mathcal{L}_\zeta$,

(b) the functions $L(\theta,\cdot)$ and $\ell(\cdot) \equiv (\zeta(\cdot))^{-1}L(\theta,\cdot)$ belong to $L^1(G)$ and $0 < \|\ell\|_1$.

Analogously we define the class $\mathcal{K}_\zeta^\psi$, for a fixed $\psi \in \Psi'$. In this case the function K satisfies an (L,ψ)-Lipschitz condition, with L satisfying (a) and (b). We simply denote by $\mathcal{K}_\zeta$ one of the classes $\mathcal{K}_\zeta^1$, $\mathcal{K}_\zeta^\psi$. When the function L is ζ-homogeneous (i.e., $R = 1$), we will write $K \in \widetilde{\mathcal{K}}_\zeta^1, \widetilde{\mathcal{K}}_\zeta^\psi$.

For $K \in \mathcal{K}_\zeta$, and the corresponding $L \in \mathcal{L}_\zeta$, we will use the following notations:

$$A_L = \|\ell\|_1 = \int_G \ell(z)\,d\mu(z) = \int_G (\zeta(z))^{-1} L(\theta, z)\,d\mu(z)$$

$$A'_L = \|L(\theta, \cdot)\|_1 = \int_G L(\theta, z)\,d\mu(z), \quad \widetilde{A}_L = \max\{A_L, A'_L\}$$

$$A_L(U) = \int_{G\setminus U} \ell(z)\,d\mu(z) = \int_{G\setminus U} (\zeta(z))^{-1} L(\theta, z)\,d\mu(z)$$

$$A'_L(U) = \int_{G\setminus U} L(\theta, z)\,d\mu(z)$$

$$r_m(s) = \sup_{1/m\le|u|\le m} \left|\frac{1}{u}\int_G K(s,t,u)\,d\mu(t) - 1\right|, \quad m = 1, 2, \ldots$$

$$r_0(s) = \sup_{m\in\mathbb{N}} r_m(s) = \sup_{u\neq 0} \left|\frac{1}{u}\int_G K(s,t,u)\,d\mu(t) - 1\right|$$

for any $U \in \mathcal{U}$.

Note that $r_m(s) \le r_0(s)$, for every $m \in \mathbb{N}$, and therefore $r_0(s) < +\infty$ implies $r_m(s) < +\infty$.

Let now $K \in \mathcal{K}_\zeta$ be fixed. We define the integral operator (of Urysohn type)

$$(Tf)(s) = \int_G K(s, t, f(t))\,d\mu(t), \quad s \in G, \tag{4.1}$$

for $f \in \operatorname{Dom} T$, i.e. for every $f \in L^0(G)$ for which $(Tf)(s)$ is well defined and measurable as a Haar integral for almost all $s \in G$.

4.2 Some estimates for *T*

Let ρ be a monotone modular on $L^0(G)$, and let $L^0_\rho(G)$ be the corresponding modular space.

We begin with the following embedding theorem in the case when the kernel function of T belongs to the class $\mathcal{K}^1_\zeta$.

Theorem 4.1. *Let ρ be a monotone, J-quasiconvex with a constant $M \ge 1$, and subbounded modular on $L^0(G)$. Let $K \in \mathcal{K}^1_\zeta$ and let*

$$S_{\rho,L} =: \int_G \ell(z) h_\rho(z)\,d\mu(z) < +\infty, \tag{4.2}$$

where h_ρ is the function from the subboundedness assumption on ρ. Then for every $f \in \operatorname{Dom} T$, such that $\zeta f \in L^0_\rho(G)$, we have for $\lambda > 0$,

$$\rho(\lambda T f) \le M\rho(\lambda MCRA_L\zeta f) + MA_L^{-1} S_{\rho,L}, \tag{4.3}$$

where C is the constant from the subboundedness assumption of ρ.

Proof. From the assumptions on K and putting $g = \zeta f$, we have

$$\begin{aligned} |(Tf)(s)| &\leq \int_G L(s,t)|f(t)|\,d\mu(t) \\ &\leq R\int_G \zeta(s+z)(\zeta(z))^{-1}L(\theta,z)|f(s+z)|\,d\mu(z) \\ &= R\int_G \ell(z)|g(s+z)|\,d\mu(z). \end{aligned}$$

Then, by the assumptions on the modular ρ, for any $\lambda > 0$ we have

$$\begin{aligned} \rho[\lambda Tf] &\leq \rho[\lambda R A_L \int_G \ell(z)|g(\cdot+z)|A_L^{-1}\,d\mu(z)] \\ &\leq A_L^{-1}M\int_G \ell(z)\rho[\lambda RMA_L Cg]\,d\mu(z) + A_L^{-1}M\int_G \ell(z)h_\rho(z)\,d\mu(z) \\ &= M\rho[\lambda RMA_L Cg] + \frac{MS_{\rho,L}}{A_L}, \end{aligned}$$

and so the assertion follows. □

Remark 4.1. If ρ is quasiconvex with the same constant $M \geq 1$ (or if μ is atomless), Theorem 4.1 implies that $Tf \in L^0_\rho(G)$ whenever $g = \zeta f \in L^0_\rho(G)$. The following is a simple example, in the linear case, which shows that we cannot replace in general the assumption $g \in L^0_\rho(G)$ by $f \in L^0_\rho(G)$.

Example 4.2. Let $G = (\mathbb{R}^+, \cdot)$ be the multiplicative group of positive real numbers, and let $\rho(f) = \|f\|_{L^1(G)}$. Thus we can take $M = C = R = 1$, $h_\rho = 0$. For every $\beta > 1$ and $\alpha \in]0,1[$, we put $K(s,t,u) = K_1(s,t)u$, where

$$K_1(s,t) = \left(\frac{t}{s}\right)^\beta \frac{1}{t^\alpha}\chi_{]0,s[}(t),$$

for $(s,t) \in G \times G$, and $u \in \mathbb{R}$. This kernel is homogeneous of degree $-\alpha$, and it is easy to show that

$$A_L = \int_0^{+\infty} z^\alpha K_1(1,z)z^{-1}\,dz < +\infty.$$

Now let us take the function $f(t) = t^\alpha$, if $0 < t < 1$, $f(t) = 0$, if $t > 1$. Then $f \in L^1(G)$, while $g(t) = \zeta(t)f(t) = t^{-\alpha}f(t)$ is not integrable in G and also $Tf \notin L^1(G)$.

It is easy to show that when $\zeta \in L^\infty(G)$, then $Tf \in L_\rho(G)$ whenever $f \in L_\rho(G)$. This happens, for example, for homogeneous functions L of degree zero of type $L(s,t) = H(s-t)$, $s,t \in G$.

Now we will state a second inequality for the operator T, when $K \in \mathcal{K}_\zeta^\psi$, $\psi \in \Psi'$. For this purpose, let ρ, η be two modulars and let $\psi \in \Psi'$. We recall that the triple $\{\rho, \psi, \eta\}$ is properly directed, if for every $\lambda \in]0,1[$ there is a constant $C_\lambda \in]0,1[$ satisfying the inequality

$$\rho[C_\lambda \psi(t, |g_t(\cdot)|)] \leq \eta[\lambda g_t(\cdot)],$$

for every family $\{g_t\}_{t\in G} \subset L^0(G)$, (see Section 1.4).

Finally, we will say that a function $g : G \times G \to \mathbb{R}_0^+$ satisfies *property* $(*)$ if $g(\cdot, z) \xrightarrow{\rho} 0$ as $z \to \theta$ in the sense of the topology in G, and there is a constant $\lambda' > 0$ such that $\rho[\lambda' g(\cdot, z)] \leq E_3$, for every $z \in G$.

We are ready to prove the following

Theorem 4.2. *Let ρ and η be monotone modulars on $L^0(G)$ such that ρ is J-quasi-convex with a constant $M \geq 1$, and η is subbounded. Let $K \in \mathcal{K}_\zeta^\psi$, with $\zeta \in L^\infty(G)$, and let us suppose that the triple $\{\rho, \psi, \eta\}$ is properly directed. Assume that*

$$S_{\eta,L} =: \int_G \ell(z) h_\eta(z)\, d\mu(z) < +\infty, \tag{4.4}$$

where h_η is the function from the subboundedness assumption on η. Let the function F from the definition of the class Ψ' satisfy condition $(*)$. Let $f \in L_\eta^0(G) \cap \operatorname{Dom} T$ and let $\lambda \in]0,1[$ be such that $\eta[\lambda C E_2 f] < +\infty$. Then there is $a > 0$, depending on λ, such that for every $\varepsilon > 0$, there is $U_\varepsilon \in \mathcal{U}$ such that

$$\rho(aTf) \leq M\eta(\lambda C E_2 f) + M A_L^{-1} S_{\eta,L} + E_3 M A_L^{-1} A_L(U_\varepsilon) + \varepsilon, \tag{4.5}$$

where the constants are previously introduced.

Proof. We can assume $\|\zeta\|_\infty \leq 1$. Moreover, let $\lambda' > 0$ be a constant such that $(*)$ is satisfied for F. With similar reasonings used in Theorem 4.1, we have

$$\begin{aligned}
|(Tf)(s)| &\leq R \int_G \ell(z)\zeta(s+z)\psi(s+z, |f(s+z)|)\, d\mu(z) \\
&\leq R \int_G \ell(z) E_1 \psi(z, E_2|f(s+z)|)\, d\mu(z) + R \int_G \ell(z) F(s,z)\, d\mu(z),
\end{aligned}$$

and so for $a > 0$ such that $2A_L MaRE_1 \leq C_\lambda$, and $2A_L MaR \leq \lambda'$ we have

$$\begin{aligned}
\rho[aTf] \leq{}& \frac{M}{A_L} \int_G \ell(z)\rho[2A_L MaRE_1 \psi(z, E_2|f(\cdot + z)|)]\, d\mu(z) \\
&+ \frac{M}{A_L} \int_G \ell(z)\rho[2A_L MaRF(\cdot, z)]\, d\mu(z) = J_1 + J_2.
\end{aligned}$$

Now, as the triple $\{\rho, \psi, \eta\}$ is properly directed, taking into account of the choice of a and of the subboundedness of η, we have

$$\begin{aligned} J_1 &\leq \frac{M}{A_L} \int_G \ell(z)\eta(\lambda E_2 f(\cdot + z))\, d\mu(z) \\ &\leq \frac{M}{A_L} \int_G \ell(z)\eta[\lambda C E_2 f]\, d\mu(z) + \frac{M}{A_L} S_{\eta,L} \\ &= M\eta[\lambda C E_2 f] + M A_L^{-1} S_{\eta,L}. \end{aligned}$$

Now we estimate J_2. Let $\varepsilon > 0$ be fixed. Then there exists $U = U_\varepsilon \in \mathcal{U}$ such that

$$\rho[\lambda' F(\cdot, z)] \leq \frac{\varepsilon}{M},$$

for every $z \in U_\varepsilon$. Hence, since $2A_L MaR < \lambda'$, we have

$$\begin{aligned} J_2 &\leq \frac{M}{A_L} \left\{ \int_U + \int_{G \setminus U} \right\} \ell(z)\rho[\lambda' F(\cdot, z)]\, d\mu(z) \\ &\leq \varepsilon + E_3 M A_L^{-1} A_L(U) \end{aligned}$$

which implies the assertion. □

Remark 4.2. As before we remark that if ρ is also quasiconvex with the same constant $M \geq 1$, (or if μ is atomless), under the assumptions of Theorem 4.2, we deduce $Tf \in L^0_\rho(G)$ whenever $f \in L^0_\eta(G)$. Note that the above inequality is simpler in the case the function ψ is independent of the parameter $t \in G$, i.e., $\psi(t, u) = \psi(u)$ for every $t \in G$, $u \in \mathbb{R}_0^+$. Indeed, in this case the Lipschitz condition on K is simpler and, consequently, the assumption on the triple $\{\rho, \psi, \eta\}$ is also simpler, and we can take $F = 0$, in property (3) of ψ. In this case, clearly, property $(*)$ is obviously satisfied for the function F.

4.3 Estimates of $\rho[Tf - f]$: case $K \in \mathcal{K}_\zeta^1$

In this section we will obtain estimates of the error of approximation of f by means of its transform Tf, under the integral operator T, defined by (4.1), in the case $K \in \mathcal{K}_\zeta^1$. To this end we need some further assumptions on the weight ζ, which defines the subhomogeneity assumption.

We will denote by $\mathcal{N}$ the class of all measurable functions $\zeta : G \to \mathbb{R}^+$ such that the following assumptions hold.

(N.1) There are constants $C', D' \geq 1$ such that

$$(\zeta(z))^{-1}\zeta(z + s) \leq C'\zeta(s) + D',$$

for every $s, z \in G$.

(N.2) For every compact set $V \subset G$ there results

$$|\zeta(s) - \zeta(z+s)| = o(1), \quad z \to \theta,$$

uniformly with respect to $s \in V$.

Example 4.3. (a) If $G = (\mathbb{R}^+, \cdot)$ we have $\zeta_\alpha \in \mathcal{N}$, where $\zeta_\alpha(t) = t^\alpha, \alpha \in \mathbb{R}$. More generally, every continuous submultiplicative function ζ on $\mathbb{R}^+$ clearly satisfies (N.1), (N.2) with $C' = D' = 1$.

(b) Taking again into consideration the group G of (a), any continuous function $\zeta : G \to \mathbb{R}_0^+$ belongs to the class $\mathcal{N}$, if it satisfies a global Δ'-condition, i.e., there exists $c > 0$ such that the following inequality holds

$$\zeta(ts) \le c\zeta(t)\zeta(s)$$

for every $s, t \in G$.

As in Section 3.1, for the given measurable function $f \in L^0(G)$, we will denote by $A_m, B_m, C_m, m = 1, 2, \dots,$ the following subsets of G:

$$\begin{aligned} A_m &= \{s \in G : |f(s)| > m\}, \\ B_m &= \{s \in G : |f(s)| < 1/m\}, \\ C_m &= G \setminus (A_m \cup B_m). \end{aligned}$$

Moreover for a subbounded modular ρ we put $h_{\rho,0} = \|h_\rho\|_\infty$.

We begin with the following theorem.

Theorem 4.3. *Let $\zeta \in \mathcal{N}$ be fixed and let $K \in \mathcal{K}_\zeta^1$. Let ρ be a monotone, J-quasi-convex and subbounded modular on $L^0(G)$. Let $f \in L^0_\rho(G) \cap \operatorname{Dom} T$, for which $g = \zeta f \in L^0_\rho(G)$, and $r_0 f \in L^0_\rho(G)$. Then for any $\lambda > 0$, for any $U \in \mathcal{U}$, compact $V \subset G$, and $m = 0, 1, \dots,$ the following inequality holds:*

$$\begin{aligned} \rho[\lambda(Tf - f)] \le{} & \frac{M}{A_L}\omega_\rho[5\lambda M R A_L g, U] \\ & + \frac{M}{A_L}A_L(U)\{2\rho[10\lambda M R A_L C g] + h_{\rho,0}\} \\ & + \rho\Big[10\lambda R f \int_U \ell(z)\tau_V(z)\,d\mu(z)\Big] + 2\rho[40\lambda R C' \widetilde{A}_L \chi_{G\setminus V} g] \\ & + \rho[40\lambda R D' \widetilde{A}_L \chi_{G\setminus V} f] + \rho[5\lambda A_L(U) g] \\ & + \rho[10\lambda R C' A'_L(U) g] + \rho[10\lambda R D' A'_L(U) f] + \mathcal{R}_m \end{aligned}$$

where the constants are previously introduced and

$$\begin{aligned}
\mathcal{R}_0 &= \rho[5\lambda r_0(\cdot) f(\cdot)] \\
\mathcal{R}_m &= M\rho[40\lambda MCRA_L\zeta f \chi_{G\setminus V}] + M\rho[40\lambda MCRA_L\zeta f \chi_{V\cap A_m}] \\
&\quad + M\rho[40\lambda MCRA_L\zeta f \chi_{V\cap B_m}] + \rho[40\lambda f(\cdot)\chi_{G\setminus V}] + \rho[40\lambda f(\cdot)\chi_{V\cap A_m}] \\
&\quad + \rho[40\lambda f(\cdot)\chi_{V\cap B_m}] + \rho[20\lambda r_m(\cdot) f(\cdot)] + 3MA_L^{-1}S_{\rho,L},
\end{aligned}$$

for $m = 1, 2, \ldots$

Proof. We can suppose, without loss of generality, that the second member of the previous inequality is finite. We first estimate the difference $Tf - f$. We have

$$\begin{aligned}
|(Tf)(s) - f(s)| &\le \int_G |K(s,t,f(t)) - K(s,t,f(s))|\, d\mu(t) \\
&\qquad + \left| \int_G K(s,t,f(s))\, d\mu(t) - f(s) \right| \\
&= I_1 + I_2.
\end{aligned}$$

We apply the Lipschitz condition on the kernel K and obtain

$$\begin{aligned}
I_1 &\le \int_G L(s,t)|f(t) - f(s)|\, d\mu(t) \\
&\le R \int_G \ell(z)\zeta(z+s)|f(z+s) - f(s)|\, d\mu(z).
\end{aligned}$$

Let U be a fixed (measurable) neighbourhood of θ. Then

$$\begin{aligned}
I_1 &\le R \int_G \ell(z)|g(z+s) - g(s)|\, d\mu(z) \\
&\quad + R|f(s)| \int_U \ell(z)|\zeta(s) - \zeta(z+s)|\, d\mu(z) \\
&\quad + R|g(s)| \int_{G\setminus U} \ell(z)\, d\mu(z) + R|f(s)| \int_{G\setminus U} \ell(z)\zeta(z+s)\, d\mu(z) \\
&= I_1^1 + I_1^2 + I_1^3 + I_1^4.
\end{aligned}$$

Let $\lambda > 0$. By the properties of the modular ρ, we have

$$\rho[\lambda(Tf - f)] \le \sum_{j=1}^{4} \rho[5\lambda I_1^j] + \rho[5\lambda I_2].$$

Step 1. *Estimation of* $\rho[5\lambda I_1^1]$.

$$\begin{aligned}\rho[5\lambda I_1^1] &\leq \frac{M}{A_L}\int_G \ell(z)\rho[5\lambda MRA_L|g(z+\cdot)-g(\cdot)|]\,d\mu(z)\\ &\leq \frac{M}{A_L}\int_U \ell(z)\rho[5\lambda MRA_L|g(z+\cdot)-g(\cdot)|]\,d\mu(z)\\ &\quad + \frac{M}{A_L}\int_{G\setminus U} \ell(z)\rho[10\lambda MRA_L|g(z+\cdot)|]\,d\mu(z)\\ &\quad + \frac{M}{A_L}\int_{G\setminus U} \ell(z)\rho[10\lambda MRA_L g]\,d\mu(z)\\ &= J_1+J_2+J_3.\end{aligned}$$

We have

$$J_1 \leq \frac{M}{A_L}\omega_\rho[5\lambda MRA_L g, U],$$

while, by subboundedness of ρ,

$$\begin{aligned}J_2 &\leq \frac{M}{A_L}\int_{G\setminus U} \ell(z)\{\rho[10\lambda MRA_L Cg]+h_{\rho,0}\}\,d\mu(z)\\ &= \{\rho[10\lambda MRA_L Cg]+h_{\rho,0}\}\frac{M}{A_L}\int_{G\setminus U}\ell(z)\,d\mu(z)\end{aligned}$$

and finally

$$J_3 = \rho[10\lambda MRA_L g]\frac{M}{A_L}\int_{G\setminus U}\ell(z)\,d\mu(z).$$

To summarize, taking into account that $C \geq 1$, we obtain the estimate of $\rho[5\lambda I_1^1]$,

$$\begin{aligned}\rho[5\lambda I_1^1] \leq{}& \frac{M}{A_L}\omega_\rho[5\lambda MRA_L g, U]\\ &+ \frac{M}{A_L}\{2\rho[10\lambda MRA_L Cg]+h_{\rho,0}\}\int_{G\setminus U}\ell(z)\,d\mu(z).\end{aligned} \tag{4.6}$$

Step 2. *Estimation of* $\rho[5\lambda I_1^2]$. Let V be a fixed compact subset of G. Putting

$$\tau_V(z) = \sup_{s\in V}|\zeta(s+z)-\zeta(s)|, \quad z\in U,$$

we may write

$$\begin{aligned}
\rho[5\lambda I_1^2] &= \rho[5\lambda|f(\cdot)|R\int_U \ell(z)|\zeta(\cdot) - \zeta(\cdot + z)|\,d\mu(z)] \\
&\leq \rho[10\lambda R\chi_V(\cdot)|f(\cdot)|\int_U \ell(z)|\zeta(\cdot) - \zeta(\cdot + z)|\,d\mu(z)] \\
&\quad + \rho[10\lambda R\chi_{G\setminus V}(\cdot)|f(\cdot)|\int_U \ell(z)|\zeta(\cdot) - \zeta(\cdot + z)|\,d\mu(z)] \\
&\leq \rho[10\lambda R|f(\cdot)|\int_U \ell(z)\tau_V(z)\,d\mu(z)] + \rho[20\lambda RA_L\chi_{G\setminus V}(\cdot)|g(\cdot)|] \\
&\quad + \rho[20\lambda R\chi_{G\setminus V}|f(\cdot)|\int_U \ell(z)\zeta(\cdot + z)\,d\mu(z)].
\end{aligned}$$

From assumption (N.1), the last term of the above inequality is dominated by

$$\rho[40\lambda RC'A'_L\chi_{G\setminus V}(\cdot)|g(\cdot)|] + \rho[40\lambda RD'A'_L\chi_{G\setminus V}(\cdot)|f(\cdot)|].$$

Thus $\rho[5\lambda I_1^2]$ is estimated by

$$\begin{aligned}
\rho[5\lambda I_1^2] &\leq \rho[10\lambda R|f(\cdot)|\int_U \ell(z)\tau_V(z)\,d\mu(z)] + \rho[20\lambda RA_L\chi_{G\setminus V}(\cdot)|g(\cdot)|] \\
&\quad + \rho[40\lambda RC'A'_L\chi_{G\setminus V}(\cdot)|g(\cdot)|] + \rho[40\lambda RD'A'_L\chi_{G\setminus V}(\cdot)|f(\cdot)|],
\end{aligned}$$

and taking into account that $C' \geq 1$, and the definition of $\widetilde{A}_L$, we finally have

$$\begin{aligned}
\rho[5\lambda I_1^2] &\leq \rho[10\lambda R|f(\cdot)|\int_U \ell(z)\tau_V(z)\,d\mu(z)] \\
&\quad + 2\rho[40\lambda RC'\widetilde{A}_L\chi_{G\setminus V}(\cdot)|g(\cdot)|] + \rho[40\lambda RD'\widetilde{A}_L\chi_{G\setminus V}(\cdot)|f(\cdot)|].
\end{aligned} \tag{4.7}$$

Step 3. *Estimation of* $\rho[5\lambda I_1^j]$, $j = 3, 4$. We have by definition

$$\rho[5\lambda I_1^3] \leq \rho[5\lambda A_L(U)|g(\cdot)|], \tag{4.8}$$

while for $\rho[5\lambda I_1^4]$ we have

$$\rho[5\lambda I_1^4] \leq \rho[10\lambda RC'A'_L(U)g] + \rho[10\lambda RD'A'_L(U)f]. \tag{4.9}$$

Step 4. *Estimation of* $\rho[5\lambda I_2]$. For $m = 0$, by definition, we easily have

$$I_2 \leq r_0(s)|f(s)|, \quad s \in G,$$

and so

$$\rho[5\lambda I_2] \leq \rho[5\lambda r_0(\cdot)f(\cdot)], \tag{4.10}$$

and the assertion follows from (4.6), (4.7), (4.8), (4.9) and (4.10) for $m = 0$.

Let us take now $m = 1, 2, \ldots$. Then

$$\begin{aligned} \rho[5\lambda I_2] &= \rho\left[5\lambda\left|\int_G K(\cdot,t,f(\cdot))\,d\mu(t) - f(\cdot)\right|\right] \\ &\le \rho\left[20\lambda\left|\int_G K(\cdot,t,f(\cdot)\chi_{G\setminus V}(\cdot))\,d\mu(t) - f(\cdot)\chi_{G\setminus V}(\cdot)\right|\right] \\ &\quad + \rho\left[20\lambda\left|\int_G K(\cdot,t,f(\cdot)\chi_{V\cap A_m}(\cdot))\,d\mu(t) - f(\cdot)\chi_{V\cap A_m}(\cdot)\right|\right] \\ &\quad + \rho\left[20\lambda\left|\int_G K(\cdot,t,f(\cdot)\chi_{V\cap B_m}(\cdot))\,d\mu(t) - f(\cdot)\chi_{V\cap B_m}(\cdot)\right|\right] \\ &\quad + \rho\left[20\lambda\left|\int_G K(\cdot,t,f(\cdot)\chi_{V\cap C_m}(\cdot))\,d\mu(t) - f(\cdot)\chi_{V\cap C_m}(\cdot)\right|\right]. \end{aligned}$$

For every measurable subset $P \subset G$, from Theorem 4.1 we have

$$\begin{aligned} &\rho\left[20\lambda\left|\int_G K(\cdot,t,f(\cdot)\chi_P(\cdot))\,d\mu(t) - f(\cdot)\chi_P(\cdot)\right|\right] \\ &\quad \le \rho\left[40\lambda\int_G |K(\cdot,t,f(\cdot)\chi_P(\cdot))|\,d\mu(t)\right] + \rho[40\lambda f(\cdot)\chi_P(\cdot)] \\ &\quad \le M\rho[40\lambda MCRA_L\zeta f\chi_P] + \frac{M}{A_L}S_{\rho,L} + \rho[40\lambda f\chi_P]. \end{aligned}$$

Thus writing the above inequality for $P = G \setminus V$, $P = V \cap A_m$ and $P = V \cap B_m$, by the definition of $r_m(s)$, we obtain the required expression for $\mathcal{R}_m$ and the assertion follows for $m = 1, 2, \ldots$. □

Remark 4.3. Note that for the validity of the inequality of Theorem 4.3 in the case $m = 1, 2, \ldots$, we can assume $r_m f \in L^0_\rho(G)$, for every m, instead of $r_0 f \in L^0_\rho(G)$ (the proof is left to the reader).

4.4 Estimates of $\rho[Tf - f]$: case $K \in \mathcal{K}^\psi_\zeta$

Here we will obtain an estimation of the modular distance between Tf and f for the kernel functions $K \in \mathcal{K}^\psi_\zeta$. In this case we have to assume that ζ is an essentially bounded function. This restriction seems to be very strong but, as an example, any convolution Mellin operator has a kernel which is homogeneous of degree zero, i.e., $\zeta(t) = 1$, for every $t \in G$. Moreover for compact groups we can take for ζ any continuous function defined on G.

If $\zeta \in L^\infty(G)$, then we do not need $\zeta \in \mathcal{N}$.

Let ρ, η be two modulars and let $\psi \in \Psi'$. Using the previous notations, we state the following

Theorem 4.4. *Let $\zeta \in L^\infty(G)$ be fixed and let $K \in \mathcal{K}_\zeta^\psi$. Let the function F from the definition of the class Ψ' satisfy $(*)$. Let ρ, η be two finite, monotone and absolutely continuous modulars on $L^0(G)$, and let us suppose that ρ is J-quasiconvex, with a constant $M \geq 1$, and η is subbounded. Assume that the triple $\{\rho, \psi, \eta\}$ is properly directed. Let $f \in L^0_{\eta+\rho}(G) \cap \mathrm{Dom}\, T$ such that $r_m f \in L^0_\rho(G)$, for every $m = 1, 2, \dots$, and let $\lambda \in]0, 1[$ be such that $\eta(2\lambda C E_2 f) < +\infty$. Then there is a constant $\nu > 0$, depending only on f and λ, such that for every $\varepsilon > 0$ there are $U = U_\varepsilon \in \mathcal{U}$ and $\overline{m} \in \mathbb{N}$ such that, for every fixed compact subset $V \subset G$, the following inequality holds:*

$$\begin{aligned}
\rho[\nu(Tf - f)] \leq\ & M\omega_\eta[\lambda E_2 f, U] \\
& + \frac{M}{A_L}\{2\eta[2\lambda C E_2 f] + h_{\eta,0}\} A_L(U) + \frac{M E_3}{A_L} A_L(U) \\
& + 7\varepsilon + 3MA_L^{-1} S_{\eta,L} + 3E_3 A_L(U_\varepsilon) + M\eta[\lambda C E_2 f \chi_{G\setminus V}] \\
& + \rho[16\nu f \chi_{G\setminus V}] + \rho[8\nu r_{\overline{m}}(\cdot) f(\cdot)],
\end{aligned}$$

where C is the constant from the subboundedness of η and $h_{\eta,0} = \|h_\eta\|_\infty$, h_η being the corresponding function.

Proof. Arguing as in the proof of Theorem 4.3, we can write

$$\begin{aligned}
|(Tf)(s) - f(s)| &\leq \int_G L(s,t)\psi(t, |f(t) - f(s)|)\, d\mu(t) \\
&\quad + \left| \int_G K(s, t, f(s))\, d\mu(t) - f(s) \right| \\
&= I_1 + I_2.
\end{aligned}$$

We first estimate I_1. Assuming that $\|\zeta\|_\infty \leq 1$, one has

$$\begin{aligned}
I_1 &\leq R \int_G \ell(z)\psi(s + z, |f(s+z) - f(s)|)\, d\mu(z) \\
&\leq RE_1 \int_G \ell(z)\psi(z, E_2|f(s+z) - f(s)|)\, d\mu(z) \\
&\quad + R \int_G \ell(z) F(s, z)\, d\mu(z) \\
&= I_1^1 + I_1^2.
\end{aligned}$$

Let $\lambda > 0$ be such that $\eta(2\lambda C E_2 f) < +\infty$, and let λ' be the constant from property $(*)$. Let $\nu > 0$ such that $16\nu R A_L M E_1 \leq C_\lambda$ and $4\nu R A_L M \nu \leq \lambda'$. From the properties of the modular ρ we have

$$\begin{aligned}
\rho[\nu(Tf - f)] &\leq \rho[2\nu I_1] + \rho[2\nu I_2] \\
&\leq \rho[4\nu I_1^1] + \rho[4\nu I_1^2] + \rho[2\nu I_2] \\
&= J_1 + J_2 + J_3,
\end{aligned}$$

and therefore

$$\begin{aligned} J_1 &\leq \frac{M}{A_L}\int_G \ell(z)\rho[4\nu R A_L M E_1 \psi(z, E_2|f(\cdot+z)-f(\cdot)|)]\,d\mu(z) \\ &\leq \frac{M}{A_L}\int_G \ell(z)\eta[\lambda E_2|f(\cdot+z)-f(\cdot)|]\,d\mu(z). \end{aligned}$$

Let $\varepsilon > 0$ and let $U = U_\varepsilon \in \mathcal{U}$ be such that $\rho[\lambda' F(\cdot, z)] < \varepsilon/M$, for every $z \in U_\varepsilon$, and $\rho[\lambda' F(\cdot, z)] \leq E_3$, for every $z \in G$. Then we have

$$\begin{aligned} J_1 &\leq M\omega_\eta[\lambda E_2 f, U] \\ &\quad + \frac{M}{A_L}\{\eta(2\lambda C E_2 f) + h_{\eta,0}\}A_L(U) + \frac{M}{A_L}\eta(2\lambda E_2 f)A_L(U), \end{aligned}$$

and thus taking into account that $C \geq 1$,

$$J_1 \leq M\omega_\eta[\lambda E_2 f, U] + \frac{M}{A_L}\{2\eta[2\lambda C E_2 f] + h_{\eta,0}\}A_L(U).$$

Next, we estimate J_2. We have

$$\begin{aligned} J_2 &= \rho[4R\nu \int_G \ell(z)F(\cdot, z)\,d\mu(z)] \\ &\leq \frac{M}{A_L}\int_G \ell(z)\rho[4RMA_L\nu F(\cdot, z)]\,d\mu(z) \\ &= \frac{M}{A_L}\left\{\int_U + \int_{G\setminus U}\right\}\ell(z)\rho[4RMA_L\nu F(\cdot, z)]\,d\mu(z) \\ &\leq \varepsilon + \frac{ME_3}{A_L}A_L(U). \end{aligned}$$

To evaluate $J_3 = \rho[2\nu I_2]$, let V be a compact subset of G. As in Theorem 4.3, we have

$$\begin{aligned} J_3 &\leq \rho\left[8\nu\left|\int_G K(\cdot, t, f(\cdot)\chi_{G\setminus V}(\cdot))\,d\mu(t) - f(\cdot)\chi_{G\setminus V}(\cdot)\right|\right] \\ &\quad + \rho\left[8\nu\left|\int_G K(\cdot, t, f(\cdot)\chi_{V\cap A_m}(\cdot))\,d\mu(t) - f(\cdot)\chi_{V\cap A_m}(\cdot)\right|\right] \\ &\quad + \rho\left[8\nu\left|\int_G K(\cdot, t, f(\cdot)\chi_{V\cap B_m}(\cdot))\,d\mu(t) - f(\cdot)\chi_{V\cap B_m}(\cdot)\right|\right] \\ &\quad + \rho\left[8\nu\left|\int_G K(\cdot, t, f(\cdot)\chi_{V\cap C_m}(\cdot))\,d\mu(t) - f(\cdot)\chi_{V\cap C_m}(\cdot)\right|\right]. \end{aligned}$$

So by using Theorem 4.2 in place of Theorem 4.1, arguing as in Theorem 4.3 and taking into account that $16\nu RA_L ME_1 < C_\lambda$, we obtain

$$\begin{aligned}
J_3 \leq\ & M\{\eta(\lambda CE_2 f\chi_{V\cap A_m}) + \eta(\lambda CE_2 f\chi_{V\cap B_m})\} + \eta[\lambda CE_2 f\chi_{G\setminus V}] + 3MA_L^{-1}S_{\eta,L} \\
& + 3E_3 A_L(U_\varepsilon) + \rho[16\nu f\chi_{G\setminus V}] + \rho[16\nu f\chi_{V\cap A_m}] + \rho[16\nu f\chi_{V\cap B_m}] \\
& + 3\varepsilon + \rho[8\nu r_m f].
\end{aligned}$$

Since $f \in L^0_{\rho+\eta}(G)$ and $|f(s)| < +\infty$, a.e. in G, by finiteness of η and ρ, and following the same reasonings as for Theorem 3.3, for the fixed ε, there exists an integer $\overline{m}$ such that

$$\begin{aligned}
&\eta[\lambda CE_2 f\chi_{V\cap A_{\overline{m}}}] < \varepsilon/2M, \quad \eta[(1/\overline{m})CE_2\lambda\chi_V] \leq \varepsilon/2M, \\
&\rho[16\nu f\chi_{V\cap A_{\overline{m}}}] < \varepsilon, \qquad\qquad \rho[(1/\overline{m})16\nu\chi_V] < \varepsilon.
\end{aligned}$$

So we have

$$\begin{aligned}
M\eta[\lambda & CE_2 f\chi_{V\cap A_{\overline{m}}}] + M\eta[\lambda CE_2 f\chi_{V\cap B_{\overline{m}}}] + \rho[16\nu f\chi_{V\cap A_{\overline{m}}}] + \rho[16\nu f\chi_{V\cap B_m}] \\
&\leq \varepsilon/2 + M\eta[(1/\overline{m})\lambda CE_2\chi_V] + \rho[16\nu f\chi_{V\cap A_{\overline{m}}}] + \rho[(1/\overline{m})16\nu\chi_V] \\
&< 3\varepsilon.
\end{aligned}$$

Thus finally we obtain

$$\begin{aligned}
J_3 \leq\ & 6\varepsilon + 3MA_L^{-1}S_{\eta,L} + 3E_3 A_L(U_\varepsilon) + M\eta[\lambda CE_2 f\chi_{G\setminus V}] \\
& + \rho[16\nu f\chi_{G\setminus V}] + \rho[8\nu r_{\overline{m}}(\cdot)f(\cdot)],
\end{aligned}$$

and thus the assertion follows. □

Remark 4.4. Note that, as in Theorem 4.2, the inequality stated in Theorem 4.4, becomes simpler if the function ψ is independent of the parameter $t \in G$. If we drop the assumption of absolute continuity and finiteness of the modulars ρ and η, we can obtain the following less sharp inequality

$$\begin{aligned}
\rho[\nu(Tf - f)] \leq\ & M\omega_\eta[\lambda E_2 f, U] \\
& + \frac{M}{A_L}\{2\eta[2\lambda CE_2 f] + h_{\eta,0}\}A_L(U) \\
& + \rho[2\nu r_0(\cdot)f(\cdot)] + \frac{ME_3}{A_L}A_L(U) + \varepsilon,
\end{aligned}$$

by involving the function $r_0(s)$ instead of $r_m(s)$ and assuming $r_0 f \in L^0_\rho(G)$.

Moreover, it is clear that the inequality given in Theorem 4.4 is also satisfied if we replace the term $\rho[8\nu r_{\overline{m}}(\cdot)f(\cdot)]$ by $\rho[8\nu r_0(\cdot)f(\cdot)]$.

4.5 Convergence theorems

In the following, we will use the Remark 2.1 (a) about the absolute continuity of a monotone modular.

Here we apply the results of the previous sections to the approximation of a function in a modular space by means of nets of nonlinear integral operators of the form (4.1). For the sake of simplicity, for a set of indices, let $W \subset \mathbb{R}^+$ be chosen as in Section 3.2, i.e. $W = [a, w_0[$ with $w_0 \in \widetilde{\mathbb{R}}_0^+$, endowed with its natural order and the notions of convergence specified in Section 3.2.

Let $\{\zeta_w\}_{w\in W}$ be a family of functions in $\mathcal{N}$ such that C', D' in (N.1) are absolute constants for any $w \in W$ sufficiently near to w_0, i.e., there exists $\overline{w} \in W$ such that C', D' are independent of $w \in]\overline{w}, w_0[$ and in (N.2) the uniformity is also taken with respect to $w \in W$, $w \in]\overline{w}, w_0[$ i.e., the family $\{\zeta_w\}_{w\in W, w>\overline{w}}$ is uniformly equicontinuous on every compact subset $V \subset G$. Moreover we will assume that the functions ζ_w are uniformly bounded by a measurable, locally bounded (i.e. bounded on every compact subset $V \subset G$) function $\zeta : G \to \mathbb{R}^+$, for $w \in]\overline{w}, w_0[$.

Let $\mathbb{K} = (K_w)_{w\in W}$, $K_w \in \mathcal{K}_{\zeta_w}$ be a family of kernel functions and for the corresponding family of functions $L_w \in \mathcal{L}_{\zeta_w}$, we will put $\ell = \ell_w$, $A_{L_w} = A_w$, $A_{L_w}(U) = A_w(U)$, $A'_{L_w}(U) = A'_w(U)$, $A'_{L_w} = A'_w$, $\widetilde{A}_{L_w} = \widetilde{A}_w$, $S_{\eta,L} = S_{w,\eta}$, $S_{\rho,L} = S_{w,\rho}$ and $r_m(s) = r_{m,w}(s)$, for every $s \in G$, $w \in W$. If the family $\{K_w\}$ satisfies the above conditions, then we will write $\{K_w\}_{w\in W} \in \mathcal{K}_{\{\zeta_w\}}$.

Thus we have a family of nonlinear operators $\mathbb{T} = \{T_w\}$, defined by

$$(T_w f)(s) = \int_G K_w(s, t, f(t))\, d\mu(t), \quad s \in G,$$

for any $f \in \operatorname{Dom} \mathbb{T} = \bigcap_{w\in W} \operatorname{Dom} T_w$.

In order to state the modular convergence theorems, we introduce a concept of *singularity* for the family of kernel functions $\{K_w\}_{w\in W}$, in a similar manner as in Section 3.2. We will say that the family $\{K_w\}_{w\in W}$, with $K_w \in \mathcal{K}_{\zeta_w}$, is a *singular kernel* if the following assumptions hold:

(j) there exist $a_1 > 0$, $a_2 \geq 1$ such that $a_1 \leq \|\ell_w\|_1 \leq a_2$ and $\widetilde{A}_w \leq a_2$, for every $w \in W$,

(jj) for every $U \in \mathcal{U}$, we have $A'_w(U) + A_w(U) \to 0$, as $w \to w_0^-$,

(jjj) for every $m = 1, 2, \ldots$, there is a positive measurable function $p_m : G \to \mathbb{R}_0^+$ such that for $w \in W$, sufficiently near to w_0, $r_{m,w}(s) \leq p_m(s)$ for every $s \in G$, and

$$\lim_{w\to w_0^-} r_{m,w}(s) = 0,$$

a.e. for $s \in G$, for every $m = 1, 2, \ldots$

Finally the family $\{K_w\}_{w\in W}$ is called *strongly singular* if in (jjj) we have

$$\lim_{w\to w_0^-} r_{0,w}(s) = 0,$$

and there is a measurable function $p : G \to \mathbb{R}_0^+$ such that $r_{0,w}(s) \le p(s)$, for every $w \in W$, sufficiently near to w_0.

It is clear that strong singularity implies singularity. The above concepts are equivalent to those introduced in Section 3.2 for kernels of type

$$K(s,t,u) = K_1(t-s,u).$$

We begin with the case of singular kernel $\{K_w\} \in \mathcal{K}^1_{\{\zeta_w\}}$ (see Section 4.1).

Theorem 4.5. *Let $\{K_w\}_{w\in W} \in \mathcal{K}^1_{\{\zeta_w\}}$ be a singular kernel. Let ρ be a monotone, absolutely finite, absolutely continuous, J-quasiconvex with a constant $M \ge 1$ and strongly subbounded modular on $L^0(G)$. Let $f \in L^0_\rho(G) \cap \operatorname{Dom}\mathbb{T}$ be such that $g = \zeta f \in L^0_\rho(G)$ and $p_m f \in L^0_\rho(G)$, for every $m = 1, 2, \ldots$. If $S_{w,\rho} \to 0$, as $w \to w_0^-$ then there exists $\lambda > 0$ such that*

$$\lim_{w\to w_0^-} \rho[\lambda(T_w f - f)] = 0. \tag{4.11}$$

Proof. Without loss of generality we can assume that there is a measurable function $p : G \to \mathbb{R}_0^+$ such that $p_m \le p$, for every $m \in \mathbb{N}$, and $g' = pf \in L^0_\rho(G)$. We will use the notations of the previous section. Putting $\zeta_w f = g_w$, we have $|g_w| \le |g|$, for $w \in W$, sufficiently near to w_0 and so we can replace g by g_w in the inequality of Theorem 4.3, written with T_w. Let $f \in L^0_\rho(G)$ such that $g, g' \in L^0_\rho(G)$. There is $a > 0$ such that $\rho(Maf) + \rho(Mag) < +\infty$ and $\rho(ag') < +\infty$. Let $0 < \alpha < 1$ be the constant from absolute continuity of ρ. Choose $\lambda > 0$ such that $40\lambda R\nu a_2 \le a\alpha$, where $\nu = \max\{C, C', D'\}$, $5\lambda M R a_2 < a$ and $20\lambda < a$. Let $\varepsilon > 0$ be fixed and we can suppose $\varepsilon < 1$. Thus for a suitable compact $V = V_\varepsilon$, we have

$$2\rho[40\lambda R\nu a_2 \chi_{G\setminus V} g] + \rho[40\lambda R\nu a_2 \chi_{G\setminus V} f] < 3\varepsilon.$$

Note that from the assumption $f \in L^\rho(G)$, it follows that

$$\lim_{m\to+\infty} \mu(V \cap A_m) = 0.$$

Thus, since the modular ρ is finite and absolutely continuous, by using the local boundedness of the function ζ, there is $\overline{m}$ such that

$$M\rho[40\lambda M\nu a_2 f \chi_{V\cap A_{\overline{m}}}] + M\rho[40\lambda M\nu a_2 g \chi_{V\cap A_{\overline{m}}}] < \varepsilon$$

and

$$\rho[40\lambda M\nu a_2 (1/\overline{m})\chi_V] + \rho[40\lambda M\nu a_2 (1/\overline{m})\zeta \chi_V] < \varepsilon.$$

So we have

$$\begin{aligned} &M\rho[40\lambda MCRa_2 g\chi_{G\setminus V}] + M\rho[40\lambda MCRa_2 g\chi_{V\cap A_{\overline{m}}}] \\ &\quad + M\rho[40\lambda MCRa_2 g\chi_{V\cap B_{\overline{m}}}] + \rho[40\lambda f(\cdot)\chi_{G\setminus V}] \\ &\quad + \rho[40\lambda f(\cdot)\chi_{V\cap A_{\overline{m}}}] + \rho[40\lambda f(\cdot)\chi_{V\cap B_{\overline{m}}}] \\ &< H\varepsilon, \end{aligned}$$

for a suitable constant $H > 0$. Let $\overline{w} = \overline{w}_\varepsilon$ be such that $\{\zeta_w\}_{w\in]\overline{w},w_0[}$ is uniformly equicontinuous on V_ε. By Theorem 2.4, we can choose the constant a, independent of ε, such that

$$\frac{M}{a_1}\omega_\rho(ag, U_\varepsilon) < \varepsilon$$

for a suitable $U_\varepsilon \in \mathcal{U}$. Thus, since $5\lambda MRa_2 \le a$, we have

$$\frac{M}{a_1}\omega_\rho[5\lambda MRa_2 g, U_\varepsilon] < \varepsilon.$$

Furthermore we can suppose that U_ε is chosen in such a way that

$$\tau_{V_\varepsilon}(z) \equiv \sup_{w\in\mathcal{W}, w>\overline{w}} \sup_{s\in V_\varepsilon} |\zeta_w(s+z) - \zeta_w(s)| < \frac{a\varepsilon}{10\lambda Ra_2},$$

for every $z \in U_\varepsilon$. Thus for any $w \in \mathcal{W}$, $w > \overline{w}$, we have

$$\rho[10\lambda Rf \int_{U_\varepsilon} \ell_w(z)\tau_{V_\varepsilon}(z)\,d\mu(z)] \le \rho[a\varepsilon f] \le M\varepsilon\rho[aMf].$$

Now, for every $w \in \mathcal{W}$, $w > \overline{w}$, we deduce

$$\begin{aligned} \rho[\lambda(T_w f - f)] \le (H+1)\varepsilon + M\varepsilon\rho[aMf] &\\ + \frac{M}{a_1}A_w(U_\varepsilon)\{2\rho[10\lambda MRa_2Cg] + h_{\rho,0}\} &\\ + \rho[5\lambda A_w(U_\varepsilon)g] + \rho[10\lambda RC'A'_w(U_\varepsilon)g] &\\ + \rho[10\lambda RD'A'_w(U_\varepsilon)f] &\\ + \rho[20\lambda r_{\overline{m},w}(\cdot)f(\cdot)] + 3\frac{M}{a_1}S_{w,\rho}. & \end{aligned} \tag{4.12}$$

Next, for the fixed $\varepsilon > 0$, we have

$$\lim_{w\to w_0^-} \{A_w(U_\varepsilon) + A'_w(U_\varepsilon)\} = 0,$$

and so since $f, g \in L^0_\rho(G)$, we obtain

$$\rho[5\lambda A_w(U_\varepsilon)g] + \rho[10\lambda RC'A'_w(U_\varepsilon)g] + \rho[10\lambda RD'A'_w(U_\varepsilon)f] \to 0, \quad w \to w_0^-.$$

Finally, since $r_{\overline{m},w}(s) \to 0$ a.e. and in view of the inequalities $r_{\overline{m},w}(s) \leq p(s)$ and $20\lambda < a$, we can apply Theorem 2.1, obtaining

$$\rho[20\lambda r_{\overline{m},w} f] \to 0, \quad w \to w_0^-$$

and so the assertion follows from (4.12) and from the assumption $S_{w,\rho} \to 0$, letting $w \to w_0^-$. □

Remark 4.5. We remark that in the case of a strongly singular kernel, the proof of the previous theorem is easier, because we only have to estimate the term $\mathcal{R}_0$ in the inequality of Theorem 4.3. This estimation is immediate taking into account of $pf \in L^0_\rho(G)$, and Theorem 2.1.

Analogously we can obtain a modular convergence theorem when $K_w \in \mathcal{K}^{\psi}_{\zeta_w}$, with $\psi \in \Psi'$, and $\{\zeta_w\}_{w \in W}$ is uniformly bounded by a constant. As in Theorem 4.4 we can assume, without loss of generality, that the constant is equal to 1. This happens for example for convolution integral operators on $\mathbb{R}^N_+$ in which we have $\zeta_w(t) = 1$, for every $t \in \mathbb{R}^N_+$, $w \in \mathcal{W}$. As in Section 4.4, in this case we do not need $\zeta_w \in \mathcal{N}$. Under this assumption on $\{\zeta_w\}$, we have the following result.

Theorem 4.6. *Let $\{K_w\}_{w \in W} \in \mathcal{K}^{\psi}_{\{\zeta_w\}}$ be a singular kernel. Let ρ, η be two monotone, absolutely continuous modulars on $L^0(G)$, and let us suppose that ρ is finite and J-quasiconvex, and η is absolutely finite and subbounded. Assume that the triple $\{\rho, \psi, \eta\}$ is properly directed, and let the function F from the definition of the class Ψ' satisfy $(*)$. If $S_{w,\eta} \to 0$, as $w \to w_0^-$ then for any $f \in L^0_{\eta+\rho}(G) \cap \operatorname{Dom} \mathbb{T}$ there is $\lambda > 0$ such that*

$$\lim_{w \to w_0^-} \rho[\lambda(T_w f - f)] = 0.$$

Proof. The proof follows by similar arguments, by applying Theorem 4.4, Theorems 2.1, 2.4 and the singularity assumption. Details are left to the reader. □

Remark 4.6. Note that in the case of strongly singular kernels the proof of Theorem 4.6 is easier, because the inequality of Theorem 4.4 is simpler.

4.6 Order of approximation in modular Lipschitz classes

As in Section 3.4, we now study the degree of modular approximation in modular Lipschitz classes, where the family $\{K_w\}$ satisfies a strong singularity assumption.

Let $\mathcal{T}$ be the class of all functions $\tau : G \to \mathbb{R}^+_0$, continuous at θ, with $\tau(t) \neq 0$ for $t \neq \theta$.

Here we recall the definition of modular Lipschitz classes, generated by a modular. Let ρ be a modular on $L^0(G)$. For a fixed $\tau \in \mathcal{T}$, we define the class

$$\mathrm{Lip}_\tau(\rho) = \{f \in L^0_\rho(G) : \exists \alpha > 0 \text{ with } \rho[\alpha|f(t+\cdot) - f(\cdot)|] = \mathcal{O}(\tau(t)),\ t \to \theta\}.$$

Let Ξ be the class introduced in Section 3.4, i.e. the class of all functions $\xi : W \to \mathbb{R}^+$, such that $\xi(w) \to 0$ as $w \to w_0^-$.

For a given $\xi \in \Xi$, we introduce the concept of ξ-singularity for a family of kernel functions $\{K_w\}_{w\in W} \in \mathcal{K}_{\{\zeta_w\}}$ in a similar manner as in Section 3.4.

We will say that $\{K_w\}_{w\in W}$ is a *strongly ξ-singular kernel* if the following assumptions hold:

(ξ.1) for any $w \in W$, $(1 + (\zeta(\cdot))^{-1})L_w(\theta, \cdot) = L_w(\theta, \cdot) + \ell_w(\cdot) \in L^1(G)$,

(ξ.2) there exist two constants a_1, a_2 such that $a_1 \le A_w = \|\ell_w\|_1 \le a_2$, for any $w \in W$ and $\widetilde{A}_w \le a_2$,

(ξ.3) for any neighbourhood U of θ

$$A'_w(U) + A_w(U) = \mathcal{O}(\xi(w)), \quad w \to w_0^-,$$

(ξ.4) for every $w \in W$

$$r_{0,w}(s) = \mathcal{O}(\xi(w)), \quad w \to w_0^-,$$

uniformly with respect to $s \in G$.

For a kernel $\{K_w\}_w \in \mathcal{K}^1_{\{\zeta_w\}}$ or $\mathcal{K}^\psi_{\{\zeta_w\}}$, $\psi \in \Psi$, we will consider the family of integral operators $\mathbb{T}$ defined by

$$(T_w f)(s) = \int_G K_w(s, t, f(t))\, d\mu(t) \tag{4.13}$$

for $f \in \mathrm{Dom}\,\mathbb{T}$.

In this section we will give two approximation theorems for $\{T_w\}$ in modular Lipschitz class $\mathrm{Lip}_\tau(\rho)$, for a given $\tau \in \mathcal{T}$.

At first we will state an approximation theorem in the case $\{K_w\} \in \mathcal{K}^1_{\{\zeta_w\}}$. To this end we will assume that the family of "weights" $\{\zeta_w\}_{w\in W} \subset \mathcal{N}$ satisfies the following assumptions:

(N$'$.1) there are (absolute) constants $C', D' \ge 1$ such that

$$(\zeta_w(z))^{-1}\zeta_w(z+s) \le C'\zeta_w(s) + D',$$

for every $s, z \in G$, $w \in W$,

(N$'$.2) the family $\{\zeta_w\}_{w\in W}$ is uniformly equicontinuous on G, in the sense that,

$$|\zeta_w(s) - \zeta_w(z+s)| = \mathcal{O}(\tau(z)), \quad z \to \theta,$$

uniformly with respect to $s \in G$ and $w \in W$,

(N$'$.3) there is a non-negative measurable function $\zeta : G \to \mathbb{R}_0^+$ such that $\zeta_w \leq \zeta$, for every $w \in W$.

Example 4.4. Let $G = (\mathbb{R}^+, \cdot)$. Let $\{\alpha_w\}$ be a net of real numbers such that $0 < a \leq \alpha_w \leq b$, for two fixed constants a, b. We can take the following family of functions $\{\zeta_w\}$,

$$\zeta_w(t) = \begin{cases} t^{\alpha_w}, & 0 < t < 1, \\ 1, & t \geq 1. \end{cases}$$

Then $\{\zeta_w\}$ satisfies the conditions (N$'$.1)–(N$'$.3)

As before, for $f \in L^0_\rho(G)$, we will put $g(t) = \zeta(t)f(t)$, $t \in G$.
We begin with the following theorem.

Theorem 4.7. *Let ρ be a quasiconvex and J-quasiconvex with a constant $M \geq 1$, monotone and subbounded modular on $L^0(G)$ with constant C and function h_ρ. Let $\tau \in \mathcal{T}$, $\xi \in \Xi$ be fixed. Let $\{\zeta\}_{w\in W}$ be a family of measurable functions with properties (N$'$.1)–(N$'$.3). Suppose that $\{K_w\}_{w\in W} \in \mathcal{K}^1_{\{\zeta_w\}}$ is a strongly ξ-singular kernel. Let us assume further that there is $U \in \mathcal{U}$ such that*

$$\int_U \ell_w(z)\tau(z)\,d\mu(z) = \mathcal{O}(\xi(w)), \quad w \to w_0^-. \tag{4.14}$$

If $f \in L^0_\rho(G) \cap \operatorname{Dom} \mathbb{T}$ is such that $g \in \operatorname{Lip}_\tau(\rho)$, then there is a $\lambda > 0$ such that

$$\rho[\lambda(T_w f - f)] = \mathcal{O}(\xi(w)), \quad w \to w_0^-.$$

Proof. From the properties of ζ and since $g = \zeta f \in \operatorname{Lip}_\tau(\rho)$, there are $\alpha > 0$ and $U \in \mathcal{U}$ such that

$$\begin{aligned} \rho[\alpha(g(z+\cdot) - g(\cdot))] &\leq B\tau(z), \\ |\zeta_w(z+s) - \zeta_w(s)| &\leq B\tau(z), \end{aligned}$$

for any $z \in U$, $s \in G$, and $w \in W$. We can assume that U also satisfies (4.14). Let $\lambda > 0$, be so small that $5\lambda MRA \leq \alpha$, $\rho[10CMRa_2\lambda g] < +\infty$, $\rho[\lambda M f] < +\infty$ and $\rho[\lambda M(|f| + |g|)] < +\infty$.

We first estimate $|T_w f - f|$. We have

$$
\begin{aligned}
|T_w f(s) - f(s)| &\le \left| \int_G K_w(s,t,f(t))\,dt - \int_G K_w(s,t,f(s))\,dt \right| \\
&\quad + \left| \int_G K_w(s,t,f(s))\,dt - f(s) \right| \\
&\le \int_G L_w(\theta,z)(\zeta_w(z))^{-1}\zeta_w(z+s)R|f(z+s)-f(s)|\,dz \\
&\quad + \left| \int_G K_w(s,t,f(s))\,dt - f(s) \right| \\
&\le \int_G L_w(\theta,z)(\zeta_w(z))^{-1}R|g(z+s)-g(s)|\,dz \\
&\quad + \int_G L_w(\theta,z)(\zeta_w(z))^{-1}R|f(s)||\zeta_w(s)-\zeta_w(z+s)|\,dz \\
&\quad + \left| \int_G K_w(s,t,f(s))\,dt - f(s) \right| \\
&\le \int_G L_w(\theta,z)(\zeta_w(z))^{-1}R|g(z+s)-g(s)|\,dz \\
&\quad + \int_U L_w(\theta,z)(\zeta_w(z))^{-1}R|f(s)||\zeta_w(s)-\zeta_w(z+s)|\,dz \\
&\quad + R|g(s)| \int_{G\setminus U} L_w(\theta,z)(\zeta_w(z))^{-1}\,dz \\
&\quad + R|f(s)| \int_{G\setminus U} L_w(\theta,z)\zeta(z+s)(\zeta_w(z))^{-1}\,dz \\
&\quad + \left| \int_G K_w(s,t,f(s))\,dt - f(s) \right|.
\end{aligned}
$$

Now, by the quasiconvexity and monotonicity of the modular ρ and by (4.14), for the above fixed λ, we obtain

$$
\begin{aligned}
\rho[\lambda(T_w f - f)] &\le \frac{M}{a_1} \int_G L_w(\theta,z)(\zeta_w(z))^{-1}\rho[5\lambda M R a_2|g(z+\cdot)-g(\cdot)|]\,dz \\
&\quad + \rho\left[5R\lambda|f(\cdot)| \int_U L_w(\theta,z)(\zeta_w(z))^{-1}|\zeta_w(\cdot)-\zeta_w(z+\cdot)|\,dz\right] \\
&\quad + \rho\left[5R\lambda|g(\cdot)| \int_{G\setminus U} L_w(\theta,z)(\zeta_w(z))^{-1}\,dz\right] + \\
&\quad + \rho\left[5R\lambda C'D'(|g(\cdot)|+|f(\cdot)|) \int_{G\setminus U} L_w(\theta,z)\,dz\right]
\end{aligned}
$$

$$+\rho\left[5\lambda\left|\int_G K_w(\cdot,t,f(\cdot))\,dt - f(\cdot)\right|\right]$$
$$= I_1 + I_2 + I_3 + I_4 + I_5.$$

To estimate I_1, we have

$$\begin{aligned} I_1 &= \frac{M}{a_1}\left\{\int_U + \int_{G\setminus U}\right\} L_w(\theta,z)(\zeta_w(z))^{-1}\rho[5R\lambda M a_2|g(z+\cdot)-g(\cdot)|]\,dz \\ &= I_1^1 + I_1^2. \end{aligned}$$

By assumption (4.14) and since $g \in \mathrm{Lip}_\tau(\rho)$ and $5R\lambda M a_2 \le \alpha$, we have

$$I_1^1 \le \frac{M}{a_1}\int_U L_w(\theta,z)(\zeta_w(z))^{-1}\tau(z)\,dz = \mathcal{O}(\xi(w)), \quad w \to w_0^-.$$

Next by τ-boundedness of ρ, we have

$$\begin{aligned} I_1^2 &\le \frac{M}{a_1}\int_{G\setminus U} L_w(\theta,z)(\zeta_w(z))^{-1}\rho[10\lambda M R a_2|g(z+\cdot)|]\,dz \\ &\quad + \frac{M}{a_1}\int_{G\setminus U} L_w(\theta,z)(\zeta_w(z))^{-1}\rho[10\lambda M R a_2 g]\,dz \\ &\le \frac{M}{a_1}\{2\rho[10CMRa_2\lambda g] + h_{\rho,0}\}\int_{G\setminus U} L_w(\theta,z)(\zeta_w(z))^{-1}\,dz \\ &= \mathcal{O}(\xi(w)), \quad w \to w_0^-. \end{aligned}$$

Now we estimate I_2. By the assumptions on $\{\zeta_w\}$ we obtain

$$\begin{aligned} I_2 &= \rho\left[5R\lambda|f(\cdot)|\int_U L_w(\theta,z)(\zeta_w(z))^{-1}|\zeta_w(\cdot)-\zeta_w(z+\cdot)|\,dz\right] \\ &\le \rho\left[5\lambda BRf\int_U L_w(\theta,z)(\zeta_w(z))^{-1}\tau(z)\,dz\right]. \end{aligned}$$

Hence, by assumption (4.14), there are $\overline{w} \in W$ and B' such that for every $w \in W$, $w \ge \overline{w}$,

$$\int_U L_w(\theta,z)(\zeta_w(z))^{-1}\tau(z)\,dz \le B'\xi(w).$$

We can assume $B = B'$. Thus we obtain the estimate

$$I_2 \le \rho[5\lambda B^2 R\xi(w) f].$$

Finally since $\xi(w) \to 0$ as $w \to w_0^-$, we can choose $\overline{w} \in W$ in such a way that $5B^2R\xi(w) < 1$, for any $w \in W$, $w > \overline{w}$, and so, from quasiconvexity of ρ, we have

$$I_2 \le 5B^2RM\xi(w)\rho[\lambda M f] = \mathcal{O}(\xi(w)), \quad w \to w_0^-.$$

By using the strong ξ-singularity, with similar reasonings, we easily obtain the estimation

$$I_3 = \rho\left[5R\lambda|g(\cdot)|\int_{G\setminus U} L_w(\theta, z)(\zeta_w(z))^{-1}\,dz\right] = \mathcal{O}(\xi(w)), \quad w \to w_0^-.$$

For I_4 we have

$$\begin{aligned} I_4 &= \rho\left[5R\lambda C'D'(|g(\cdot)| + |f(\cdot)|)\int_{G\setminus U} L_w(\theta, z)\,dz\right] \\ &\le \rho[5R\lambda C'D'P\xi(w)(|f| + |g|)], \end{aligned}$$

for a suitable constant $P > 0$. Thus, as in the estimations of I_2 and I_3, we get

$$I_4 = \mathcal{O}(\xi(w)), \quad w \to w_0^-.$$

In order to estimate the term

$$I_5 = \rho\left[5\lambda\left|\int_G K_w(\cdot, t, f(\cdot))\,dt - f(\cdot)\right|\right],$$

we remark that by the ξ-singularity, we have

$$\left|\int_G K_w(s, t, f(s))\,dt - f(s)\right| \le r_w(s)|f(s)|,$$

and since $r_{0,w}(s) = \mathcal{O}(\xi(w))$, uniformly with respect to $s \in G$, we easily deduce

$$I_5 = \mathcal{O}(\xi(w)), \quad w \to w_0^-.$$

Thus, taking into account of the previous estimates, we finally obtain

$$\rho[\lambda(T_w f - f)] = \mathcal{O}(\xi(w)), \quad w \to w_0^-,$$

and so the assertion follows. □

Now we state an analogous approximation theorem for the family of operators $\mathbb{T} = \{T_w\}_{w\in W}$, when the kernel $\{K_w\}_w$ belongs to the class $\mathcal{K}^{\psi}_{\{\zeta_w\}}$, where $\zeta_w : G \to \mathbb{R}^+$ is a family of essentially bounded functions on G. In this case we do not need the assumptions (N′.1)–(N′.2), while we will suppose that (N′.3) is still satisfied with an essentially bounded function ζ.

We have the following

Theorem 4.8. *Let $\psi \in \Psi'$, and let ρ, η be two modulars on $L^0(G)$ such that ρ is quasiconvex and monotone and η is strongly τ-bounded and monotone. Let us suppose that the triple $\{\rho, \psi, \eta\}$ is properly directed. Let $\tau \in \mathcal{T}, \xi \in \Xi$ be fixed* *and let* $\zeta_w : G \to \mathbb{R}^+$ *be a family of essentially bounded functions satisfying* (N$'$.3). *Suppose that* $\{K_w\}_{w\in W} \in \mathcal{K}^{\psi}_{\{\zeta_w\}}$ *is a strongly* ξ*-singular kernel. Let us assume further that there is* $U \in \mathcal{U}$ *such that*

$$\int_U \ell_w(z)\tau(z)\,d\mu(z) = \mathcal{O}(\xi(w)), \quad w \to w_0^-, \tag{4.15}$$

and that the function F *from the definition of the class* Ψ' *satisfies condition* $(*)$ *of Section* 4.2 *and there is* $\lambda' > 0$ *such that* $\rho[\lambda' F(\cdot, z)] = \mathcal{O}(\tau(z))$, *as* $z \to \theta$. *If* $f \in L^0_{\rho+\eta}(G) \cap \operatorname{Dom}\mathbb{T}$, *is such that* $f \in \operatorname{Lip}_\tau(\eta)$, *then there is* $\nu > 0$ *such that*

$$\rho[\nu(T_w f - f)] = \mathcal{O}(\xi(w)), \quad w \to w_0^-.$$

Proof. Without loss of generality, we can assume that $\|\zeta_w\|_\infty \le 1$, for every $w \in W$.

By assumptions, there are $\alpha > 0$, $B, B' > 0$ and $U \in \mathcal{U}$ such that

$$\begin{aligned} \eta[\alpha(f(z+\cdot) - f(\cdot))] &\le B\tau(z), \\ \rho[\lambda' F(\cdot, z)] &\le B'\tau(z) \end{aligned}$$

for any $z \in U$. Let $\lambda \in]0, 1[$ be fixed in such a way that $\eta(2\lambda E_2 f) < +\infty$ and $\lambda E_2 < \alpha$. Let $\nu > 0$ such that $3\nu < \lambda$, $3MRE_1\nu a_2 < C_\lambda$, and $3MRa_2 < \lambda'$, where C_λ is the constant corresponding to λ in the definition of properly directed triple. As in Theorem 4.7, we have

$$\begin{aligned} |T_w f(s) - f(s)| &\le \int_G \ell_w(z))R\psi(z+s, |f(z+s) - f(s)|)\,d\mu(z) \\ &\quad + \left|\int_G K_w(s, t, f(s))\,d\mu(t) - f(s)\right| \\ &\le \int_G \ell_w(z)RE_1\psi(z, E_2|f(z+s) - f(s)|)\,d\mu(z) \\ &\quad + R\int_G \ell_w(z)F(s, z)\,d\mu(z) \\ &\quad + \left|\int_G K_w(s, t, f(s))\,d\mu(t) - f(s)\right|. \end{aligned}$$

So, applying the modular ρ and the property of the triple $\{\rho, \eta, \psi\}$, we have

$$\begin{aligned} \rho[\nu(T_w f - f)] &\le \frac{M}{a_1}\int_G \ell_w(z)\eta[\lambda|f(z+\cdot) - f(\cdot)|)]\,d\mu(z) \\ &\quad + \frac{M}{a_1}\int_G \ell_w(z)\rho[\lambda' F(\cdot, z)]\,d\mu(z) \\ &\quad + \rho\left[3\nu\left|\int_G K_w(\cdot, t, f(\cdot))\,d\mu(t) - f(\cdot)\right|\right] \\ &= J_1 + J_2 + J_3. \end{aligned}$$

Now J_1 and J_3 can be estimated as in Theorem 4.7, while for J_2 we have

$$J_2 \leq \frac{M}{a_1}\left\{B'\int_U \ell_w(z)\tau(z)\,d\mu(z) + E_3\int_{G\setminus U}\ell_w(z)\,d\mu(z)\right\} = \mathcal{O}(\xi(w)), \quad w \to w_0^-.$$

Hence the assertion follows. □

In the following section we apply the theory developed to a particular case: the nonlinear Mellin-type operators.

4.7 Application to nonlinear weighted Mellin convolution operators

Here we consider $G = (\mathbb{R}^+, \cdot)$, equipped with the Haar measure $d\mu(t) = 1/t$.

We will consider a neighbourhood base of the neutral element 1, defined by $U_\delta =]1-\delta, 1+\delta[$, for $\delta \in]0, 1[$ and we will put $U_\delta^c = \mathbb{R}^+\setminus U_\delta$. Let $W \subset \mathbb{R}^+$ be any half-line of type $[a, +\infty[$, considered with its natural order. Let $\{H_w\}_{w\in W}$, $H_w : \mathbb{R}^+ \to \mathbb{R}_0^+$ be measurable functions, and $\{\Gamma_w\}_{w\in W}$, $\Gamma_w : \mathbb{R} \to \mathbb{R}$ be equi-lipschitzian functions (and we will assume, for the sake of simplicity, that the absolute Lipschitz constant is equal to 1) such that $\Gamma_w(0) = 0$, for every $w \in W$. Let $\{\alpha_w\}_{w\in W}$ be a net of real numbers such that $\alpha_w \to 0$, as $w \to +\infty$, and we can assume that $\alpha \leq \alpha_w \leq \beta$, for $w \in W$, where $\alpha, \beta \in \mathbb{R}$, $\alpha \leq 0 \leq \beta$. Then the family of functions $\{\zeta_w\}$, with $\zeta_w(t) = t^{\alpha_w}$, is uniformly equicontinuous on every compact set $V \subset \mathbb{R}^+$ and satisfies (N′.1) uniformly with respect to $w \in W$.

The functions $L_w : \mathbb{R}^+ \times \mathbb{R}^+ \to \mathbb{R}_0^+$, defined by

$$L_w(s,t) = t^{\alpha_w} H_w(ts^{-1})$$

are homogeneous of degree α_w, i.e. $\zeta_w(t) = t^{\alpha_w}$. In this particular case, we have

$$\begin{aligned} \ell_w(z) &= z^{-\alpha_w} L_w(1,z) = H_w(z), \quad z \in \mathbb{R}^+ \\ A_w &= \|\ell_w\|_1 = \|H_w\|_1, \\ A'_w &= \|(\cdot)^{\alpha_w} H_w(\cdot)\|_1, \\ A_w(\delta) &= \int_{U_\delta^c} H_w(z)\frac{dz}{z}, \\ A'_w(\delta) &= \int_{U_\delta^c} z^{\alpha_w} H_w(z)\,\frac{dz}{z}. \end{aligned}$$

For the sake of simplicity we limit ourselves to strongly singular kernels. We will define

$$M_w(s,t,u) = t^{\alpha_w} H_w(ts^{-1})\Gamma_w(u),$$

for $s, t \in \mathbb{R}^+$, $u \in \mathbb{R}$. Then M_w is a $(L_w, 1)$-Lipschitz function, and we will assume $\{M_w\} \in \mathcal{K}^1_{\{\zeta_w\}}$, to be a strongly singular kernel, and we will put $r_{0,w}(s) = r_w(s)$.

In particular we have

$$r_w(s) = \sup_{u \neq 0} \left| s^{\alpha_w} \frac{\Gamma_w(u)}{u} \int_0^{+\infty} z^{\alpha_w} H_w(z) \frac{dz}{z} - 1 \right|.$$

Now we consider some conditions in order that (jjj) of Section 4.5 is satisfied. Let us suppose that (j), (jj) of Section 4.5 hold. Then in particular $A'_w \leq a_2$, for sufficiently large $w \in W$. Moreover we have

$$\sup_{w \in W} \sup_{u \neq 0} \left| \frac{\Gamma_w(u)}{u} \right| \leq 1.$$

Then we can take $p(t) = 1 + a_2 \sup_{w \in W} s^{\alpha_w}$, and

$$\lim_{w \to +\infty} \sup_{u \neq 0} \left| \frac{\Gamma_w(u)}{u} A'_w s^{\alpha_w} - 1 \right| = 0,$$

for almost all $s \in \mathbb{R}^+$. For example, this happens when $\Gamma_w(u) \to u$, uniformly on $\mathbb{R}$ and $A'_w \to 1$, as $w \to +\infty$.

For the given kernel $\{M_w\}$ we define the family $\mathbb{M}$ of integral operators via

$$(\mathcal{M}_w f)(s) = \int_0^{+\infty} t^{\alpha_w} H_w(ts^{-1}) \Gamma_w(f(t)) \frac{dt}{t}, \quad s \in \mathbb{R}^+. \tag{4.16}$$

For any $w \in W$, we call the family $\mathbb{M}$ a *nonlinear Mellin convolution operator with weight* t^{α_w} or *nonlinear Mellin convolution operator of order* α_w.

Taking $\alpha_w = 0$, for every $w \in W$, we obtain the family $\overline{\mathbb{M}}$ of *nonlinear Mellin convolution operators*, defined by

$$(\overline{\mathcal{M}}_w f)(s) = \int_0^{+\infty} H_w(ts^{-1}) \Gamma_w(f(t)) \frac{dt}{t},$$

for $s \in \mathbb{R}^+$. The functions $L_w(s,t) = H_w(ts^{-1})$, $w \in W$, are homogeneous of degree 0, i.e., $\zeta_w(t) = 1$ for every $w \in W$, $t \in \mathbb{R}^+$.

Example 4.5. For every $n \in \mathbb{N}$, let us consider the function

$$(n+1) \frac{t^{n+1+\alpha_n}}{s^{n+1}} \chi_{]0,s[}(t), \quad s, t \in \mathbb{R}^+,$$

where $\{\alpha_n\}$ is a sequence of real numbers such that $\alpha_n \to 0$. Then the corresponding family of operators $\mathbb{M}$, given by

$$(\mathcal{M}_n f)(s) = (n+1) \int_0^s \frac{t^{n+1+\alpha_n}}{s^{n+1}} \Gamma_n(f(t)) \frac{dt}{t},$$

is called the *nonlinear weighted moment operator with weights* t^{α_n}, and taking $\alpha_n \equiv 0$ for any n, we obtain the *nonlinear moment operators*.

Thus the previous theory can be applied to the families $\mathbb{M} = \{\mathcal{M}\}_w$, $\overline{\mathbb{M}} = \{\overline{\mathcal{M}}\}_w$, $w \in W$, in modular function spaces, in which the modular satisfies the conditions of Section 4.5. In particular, just as an example, we can consider the Orlicz space $L^{\varphi}_{\mu}(\mathbb{R}^+)$, for a fixed φ-function $\varphi \in \Phi$, (see Example 1.5 (b)), in which the modular functional takes now the form

$$I_{\varphi}(f) = \int_{\mathbb{R}^+} \varphi(|f(t)|)t^{-1}\, dt,$$

for $f \in L^0(\mathbb{R}^+)$. We assume that the generating function $\varphi : \mathbb{R}^+_0 \to \mathbb{R}^+_0$ is J-quasi-convex, with a constant $M \geq 1$. For completeness we report the assumptions on φ.

(φ.1) φ is continuous and nondecreasing,

(φ.2) $\varphi(0) = 0$, $\varphi(u) > 0$, for $u > 0$, and $\lim_{u \to +\infty} \varphi(u) = +\infty$,

(φ.3) there is $M \geq 1$, such that, for any $g \in L^1_{\mu}(\mathbb{R}^+)$ with $\|g\|_1 = 1$ and $g \geq 0$,

$$\varphi\left(\int_0^{+\infty} |f(t)|g(t)t^{-1}\, dt\right) \leq M \int_0^{+\infty} \varphi(M|f(t)|)g(t)t^{-1}\, dt,$$

for any $f \in L^0(\mathbb{R}^+)$.

If φ satisfies ($\varphi.i$), $i = 1, 2, 3$, we will write $\varphi \in \Phi$. We will denote by $L^{\varphi}_{\mu}(\mathbb{R}^+)$ the Orlicz space generated by φ, with respect to the Haar measure $d\mu(t) = dt/t$, i.e.,

$$L^{\varphi}_{\mu}(\mathbb{R}^+) = \{f \in L^0(\mathbb{R}^+) : I_{\varphi}(\lambda f) < +\infty \text{ for some } \lambda > 0\}.$$

As a consequence of Theorem 4.1, under the previous assumptions on the functions H_w, Γ_w, we have $L^{\varphi}_{\mu}(\mathbb{R}^+) \subset \operatorname{Dom} \mathbb{M}$, for any $w \in W$ (the reader is encouraged to find the proof). We have the following corollary.

Corollary 4.1. *Let $\varphi \in \Phi$ and let us assume that the functions M_w satisfy the above conditions of singularity. Let $f \in L^{\varphi}_{\mu}(\mathbb{R}^+)$ be such that $pf \in L^{\varphi}_{\mu}(\mathbb{R}^+)$ and $\zeta f \in L^{\varphi}_{\mu}(\mathbb{R}^+)$, where $\zeta = \sup_{w \in W} s^{\alpha_w}$. Then there is $\lambda > 0$ such that*

$$\lim_{w \to +\infty} I_{\varphi}[\lambda(\mathcal{M}_w f - f)] = 0.$$

In the same manner we can obtain a version of Theorem 4.6, for the family $\overline{\mathbb{M}}$ of operators $\overline{\mathcal{M}}_w$, where the functions Γ_w satisfy a (L_w, ψ)-Lipschitz condition, in the case of Orlicz spaces generated by two φ-functions φ_1, φ_2 such that the triple $\{I_{\varphi_1}, \psi, I_{\varphi_2}\}$ is properly directed. In [34] there are given some conditions on φ_1, φ_2 in order that the above triple is properly directed. In this instance, it is clear that the function ψ is taken independent of $t \in G$.

Remark 4.7. Let us consider now the linear case, i.e. $\Gamma_w(u) = u$, for every $w \in W$. In [15] there are given some results about modular convergence of these operators, when the homogeneity of the kernel is taken with respect to the same function ζ, for every $w \in \mathcal{W}$. But in this case we obtain $\mathcal{M}_w f \to \zeta f$ in modular sense and in general we cannot replace ζf with f. Indeed if for example $\alpha_w = \alpha$, for every $w \in W$, condition (jjj) of the singularity assumption is satisfied only if $\alpha = 0$, i.e., $\zeta(t) = 1$, for every $t \in \mathbb{R}^+$. However taking a net of "weights" of type $\zeta_w(t) = t^{\alpha_w}$, such that $\alpha_w \to 0$, we can improve the result given in [15], obtaining an approximation theorem for f.

4.8 Bibliographical notes

The study of linear integral operators with homogeneous kernel (of degree α) takes its origin in the theory of inequalities. The classical Hardy, Hardy–Littlewood and Schur inequalities involve some special linear integral operator acting on L^p-spaces. Extensions of these inequalities are obtained by several authors. We quote here Flett [100], E. R. Love [136], [137], F. Feher [97], P. L. Butzer and F. Feher [58], C. Bardaro and G. Vinti [29], I. Mantellini and G. Vinti [142]. Operators with homogeneous kernels are also considered in connection with fractional calculus by A. Erdélyi in [96], H. Kober [128], K. B. Oldham and J. Spanier [169]. Also, the moment kernel may be considered as a function with degree of homogeneity zero if we consider a Haar measure, and of degree -1 if we consider the Lebesgue measure on $\mathbb{R}^+$. Results on moment type operators, in connection with approximation theory, fractional calculus and calculus of variations, are given by many authors. We quote the contribution of C. Vinti [198], who obtained, for a general class of linear operators, some convergence results with respect to some integral functional of Calculus of Variations. These results were again taken into consideration by F. Degani Cattelani [84], [85], for convergence in perimeter, and by C. Fiocchi [98] for convergence in area. Moment-type operators in approximation theory in L^p-spaces were also considered by F. Barbieri [8] and C. Bardaro [9]. In connection with fractional calculus, we recall here the contribution of Zanelli [212] and C. Fiocchi [99], in which an application of the moment operator theory to Hausdorff dimension of some fractal set is studied.

The general definition of homogeneity given in Section 4.1 takes its origin in the paper [31], in which some estimates of linear integral operators with general assumptions of homogeneity are considered with respect to some variational functional. Inequalities for such operators are also given in Musielak–Orlicz spaces in [32].

Approximation theorems in function spaces by means of general linear integral operators of type

$$\int_G K_w(s,t) f(t)\, d\mu(t),$$

with ζ-homogeneous kernel $\{K_w\}$, were given firstly in [35], in Musielak–Orlicz spaces, and then in [15] for general modular spaces.

The results given in this chapter for nonlinear operators were obtained in [43]. For Orlicz spaces some convergence results, for nonlinear integral operators with homogeneous kernels (in the sense given in Section 4.1), were given in [37]. Here, as an application, a nonlinear version of Mellin convolution integral operator is introduced. The definition of these operators is also further generalized by introducing a weight function (given essentially by the function ζ from the homogeneity assumption).

Classical Mellin (linear) convolution operators came from the theory of the Mellin transform, extensively studied by P. L. Butzer and S. Jansche [61], [62], [64], [65], in which the foundations of Mellin approximation theory are given.

The Urysohn-type operators considered here are related to convolution operators by means of the homogeneity assumption on the kernel, in the sense that these assumptions represent a generalization of convolution. By considering a suitable definition of singularity and by stating a density theorem in modular topology, in [13], [14], C. Bardaro and I. Mantellini obtained convergence theorems in Orlicz or Musielak–Orlicz spaces for families of general Urysohn-type operators, without any homogeneity assumption.

Finally we wish to point out that the assumption of homogeneity given here is a useful tool in studying approximation theory by means of operators with homogeneous kernels in abstract topological groups. The definition proposed here is different from that used in abstract harmonic analysis (in homogeneous group, see e.g. Folland–Stein [102]).

Chapter 5

Summability methods by convolution-type operators

5.1 An estimate of $\rho[\alpha(T_1 f - T_2 g)]$

In Chapter 3 we were concerned with the problem of approximation of a function $f \in L^0_{\rho+\eta}(\Omega) \cap \operatorname{Dom} \mathbb{T}$ by means of a family $\mathbb{T} = (T_w)_{w \in W}$ of convolution-type operators in the sense of ρ-convergence in a space $L^0_\rho(\Omega)$ (Theorem 3.3). Here, we shall apply the same operators to summability problem of a family $(f_w)_{w \in W}$ of functions, i.e. to the investigation of the problem of convergence of the transformed sequence $(T_w f_w)_{w \in W}$. The main problem is: under what conditions, a convergent family $(f_w)_{w \in W}$ is transformed in a convergent family $(T_w f_w)_{w \in W}$. Here, convergence is meant in the sense of the respective modulars.

In this chapter we will deal with a measure space (Ω, Σ, μ) with a σ-finite and complete measure μ, where Ω is equipped with a commutative operation $+ : \Omega \times \Omega \to \Omega$.

First of all, we need a fundamental inequality similar to the inequality (3.4) in Chapter 3.

Let $K_1 : \Omega \to \mathbb{R}$ and $K_2 : \Omega \to \mathbb{R}$ be two Carathéodory kernel functions and let $0 < L_1 \in L^1(\Omega)$, $0 < L_2 \in L^1(\Omega)$; let $\psi : \Omega \times \mathbb{R}_0^+ \to \mathbb{R}_0^+$, be a function belonging to the class Ψ (see Section 3.1), defining the Lipschitz condition. We consider the respective convolution-type operators

$$(T_1 f)(s) = \int_\Omega K_1(t, f(t+s))\, d\mu(t), \quad (T_2 g)(s) = \int_\Omega K_2(t, g(t+s))\, d\mu(t)$$

for $f \in \operatorname{Dom} T_1$ and $g \in \operatorname{Dom} T_2$. Let us write $p_1(t) = L_1(t)/\|L_1\|_1$, $p_2(t) = L_2(t)/\|L_2\|_1$ for $t \in \Omega$ and let us put $L_{1,2} = \max(\|L_1\|)_1, \|L_2\|_1)$.

Theorem 5.1. *Let ρ be a monotone, J-quasiconvex with a constant $M \geq 1$ modular on $L^0(\Omega)$ and let η be a modular on $L^0(\Omega)$, subbounded with respect to the operation $+$ with a constant $C \geq 1$ and a function $\ell \in L^\infty(\Omega)$. Let K_1 be an (L_1, ψ)-Lipschitz Carathéodory kernel function and let K_2 be an (L_2, ψ)-Lipschitz Carathéodory kernel function such that $\{\rho, \psi, \eta\}$ is a properly directed triple. Finally, let $U \in \Sigma$, $0 < \lambda < 1$ and $0 < \alpha < C_\lambda/(6ML_i)$, $i = 1, 2$, be arbitrary. Then for every*

$f, g \in L^0_\eta(\Omega) \cap \operatorname{Dom} T_1 \cap \operatorname{Dom} T_2$ *there holds the inequality*

$$\begin{aligned}\rho[\alpha(T_1 f - T_2 g)] \leq \rho\left[6\alpha \left|\int_U (K_1(t, f(t+\cdot)) - K_2(t, f(t+\cdot)))\, d\mu(t)\right|\right] \\ + M[\eta(\lambda C f) \\ + \|\ell\|_\infty]\left(\int_{\Omega\setminus U} p_1(t)\, d\mu(t) + \int_{\Omega\setminus U} p_2(t)\, d\mu(t)\right) \\ + M[\eta(\lambda C(f-g)) + \|\ell\chi_U\|_\infty] \\ + M[\eta(2\lambda C f) + \eta(2\lambda C g) + 2\|\ell\|_\infty] \int_{\Omega\setminus U} p_2(t)\, d\mu(t).\end{aligned} \tag{5.1}$$

Proof. Since

$$\begin{aligned}|(T_1 f)(s) - (T_2 g)(s)| \leq \left|\int_\Omega (K_1(t, f(t+s)) - K_2(t, f(t+s)))\, d\mu(t)\right| \\ + \left|\int_\Omega (K_2(t, f(t+s)) - K_2(t, g(t+s)))\, d\mu(t)\right|,\end{aligned}$$

so by monotonicity of ρ, we have for arbitrary $\alpha > 0$

$$\begin{aligned}\rho[\alpha(T_1 f - T_2 g)] \leq \rho\left[2\alpha \left|\int_\Omega (K_1(t, f(t+\cdot)) - K_2(t, f(t+\cdot)))\, d\mu(t)\right|\right] \\ + \rho\left[2\alpha \left|\int_\Omega (K_2(t, f(t+\cdot)) - K_2(t, g(t+\cdot)))\, d\mu(t)\right|\right].\end{aligned}$$

Denoting the two terms on the right-hand side of the last inequality by I and II, respectively, we have

$$\rho[\alpha(T_1 f - T_2 g)] \leq I + II.$$

We shall estimate I and II separately. Since ρ is a modular, we have

$$\rho(f_1 + f_2 + f_3) \leq \rho(3f_1) + \rho(3f_2) + \rho(3f_3)$$

for $f_1, f_2, f_3 \in L^0(\Omega)$. Hence by monotonicity of ρ we obtain for every $U \in \Sigma$

$$\begin{aligned}I \leq \rho\left[6\alpha \left|\int_U (K_1(t, f(t+\cdot)) - K_2(t, f(t+\cdot)))\, d\mu(t)\right|\right] \\ + \rho\left[6\alpha \int_{\Omega\setminus U} |K_1(t, f(t+\cdot))| d\mu(t)\right] + \rho\left[6\alpha \int_{\Omega\setminus U} |K_2(t, f(t+\cdot))|\, d\mu(t)\right].\end{aligned}$$

Denoting the three terms on the right-hand side of the last inequality by I', II' and III', respectively, we have

$$I \leq I' + II' + III'.$$

We shall estimate II'. Since K_1 is an (L_1, ψ)-Lipschitz kernel function, by J-quasiconvexity with constant $M \geq 1$ of ρ, we obtain

$$II' \leq M \int_{\Omega \setminus U} p_1(t)\rho[6\alpha M \|L_1\|_1 \psi(t, |f(t+\cdot)|)]\, d\mu(t).$$

Applying the assumption that $\{\rho, \psi, \eta\}$ is a properly directed triple with any $\lambda \in]0, 1[$ and the respective $C_\lambda \in]0, 1[$ we obtain for $0 < 6\alpha M \|L_1\|_1 \leq C_\lambda$

$$II' \leq M \int_{\Omega \setminus U} p_1(t)\eta[\lambda |f(t+\cdot)|]\, d\mu(t).$$

Since η is subbounded with constant $C \geq 1$ and function $\ell \in L^\infty(\Omega)$, we get

$$II' \leq M[\eta(\lambda C f) + \|\ell\|_\infty] \int_{\Omega \setminus U} p_1(t)\, d\mu(t).$$

Similarly, we get for $0 < 6\alpha M \|L_2\|_1 \leq C_\lambda$ the inequality

$$III' \leq M[\eta(\lambda C f) + \|\ell\|_\infty] \int_{\Omega \setminus U} p_2(t)\, d\mu(t).$$

Consequently, we obtain the estimation

$$\begin{aligned} I \leq \rho &\left[6\alpha \left| \int_U (K_1(t, f(t+\cdot)) - K_2(t, f(t+\cdot)))\, d\mu(t) \right| \right] \\ &+ M[\eta(\lambda C f) + \|\ell\|_\infty] \left(\int_{\Omega \setminus U} p_1(t)\, d\mu(t) + \int_{\Omega \setminus U} p_2(t)\, d\mu(t) \right). \end{aligned}$$

Now, we are going to estimate II. Since ρ is monotone and J-quasiconvex with constant $M \geq 1$, and K_2 is (L_2, ψ)-Lipschitz, we obtain

$$\begin{aligned} II &\leq \rho\left[2\alpha \int_\Omega |K_2(t, f(t+\cdot)) - K_2(t, g(t+\cdot))|\, d\mu(t) \right] \\ &\leq M \int_\Omega p_2(t)\rho[2\alpha M \|L_2\|_1 \psi(t, |f(t+\cdot) - g(t+\cdot)|)]\, d\mu(t). \end{aligned}$$

Since $\{\rho, \psi, \eta\}$ is a properly directed triple, so taking for a given $\lambda \in]0, 1[$ the number $\alpha > 0$ so small that $0 < 2\alpha M \|L_2\|_1 \leq C_\lambda$, we obtain

$$II \leq M \int_\Omega p_2(t)\eta[\lambda(f(t+\cdot) - g(t+\cdot))]\, d\mu(t).$$

Let us put

$$\nu(A) = \int_A p_2(t)\eta[\lambda(f(t+\cdot) - g(t+\cdot))]\, d\mu(t)$$

for arbitrary $A \in \Sigma$. Then we have for arbitrary $U \in \Sigma$

$$II \leq M\nu(U) + M\nu(\Omega \setminus U).$$

Since η is subbounded, we obtain

$$\nu(U) \leq \eta[\lambda C(f-g)] + \|\ell\chi_U\|_\infty$$

and

$$\begin{aligned}\nu(\Omega \setminus U) &\leq \int_{\Omega\setminus U} p_2(t)\eta(2\lambda f(t+\cdot))\,d\mu(t) + \int_{\Omega\setminus U} p_2(t)\eta(2\lambda g(t+\cdot))\,d\mu(t)\\ &\leq [\eta(2\lambda Cf) + \eta(2\lambda Cg) + 2\|\ell\|_\infty]\int_{\Omega\setminus U} p_2(t)\,d\mu(t).\end{aligned}$$

Then

$$\begin{aligned}II \leq{}& M[\eta(\lambda C(f-g)) + \|\ell\chi_U\|_\infty]\\ &+ M[\eta(2\lambda Cf) + \eta(2\lambda Cg) + 2\|\ell\|_\infty]\int_{\Omega\setminus U} p_2(t)\,d\mu(t).\end{aligned}$$

Estimations of I and II give together the inequality (5.1). □

5.2 Conservative summability methods

Let W be an infinite set of indices defined as in Section 3.2, i.e., $W \subset [a, w_0[$, where $a \in \mathbb{R}$ and $w_0 \in \mathbb{R}$, $w_0 > a$, or $w_0 = +\infty$ and w_0 is a point of accumulation of W. Convergence with respect to W is defined as left-hand side convergence $w \to w_0^-$, $w \in W$. In Section 1.2 we defined ρ-convergence, i.e., modular convergence with respect to a modular ρ on $\mathcal{X}$ for a sequence (f_n) of functions $f_n \in \mathcal{X}_\rho$ by the condition $\rho(\lambda(f_n - f)) \to 0$ as $n \to +\infty$ for some $\lambda > 0$. This definition may be transferred to the case of convergence $\overset{w}{\to}$, immediately. We say that a family $(f_w)_{w\in W}$ of functions $f_n \in \mathcal{X}_\rho$ is *ρ-convergent to a function* $f \in \mathcal{X}_\rho$, if there exists a $\lambda > 0$ (depending on the family $(f_w)_{w\in W}$) such that $\rho(\lambda(f_w - f)) \overset{w}{\to} 0$. In the same manner one may define $\|\cdot\|_\rho$-norm convergence of $(f_w)_{w\in W}$ to f by means of the condition $\|f_w - f\|_\rho \overset{w}{\to} 0$. Similarly as in the case of Theorem 1.2 it is easily proved that $\|f_w - f\|_\rho \overset{w}{\to} 0$ if and only if there holds $\rho(\lambda(f_w - f)) \overset{w}{\to} 0$, for every $\lambda > 0$. Obviously the same holds if we replace the norm $\|\cdot\|_\rho$ generated by a convex modular ρ, by the F-norm $|||\cdot|||_\rho$ generated by a general modular ρ. Besides ρ-convergence and $\|\cdot\|_\rho$-norm (or $|||\cdot|||_\rho$ $-$ F-norm) convergence, one may also introduce the ρ-Cauchy condition and the $\|\cdot\|_\rho$-Cauchy condition (or $|||\cdot|||_\rho$-Cauchy condition) for families $(f_w)_{w\in W}$. We say that the family $(f_w)_{w\in W}$ with $f_w \in \mathcal{X}_\rho$ for all $w \in W$ is *ρ-Cauchy* or satisfies the *ρ-Cauchy condition*, if there exists a $\lambda > 0$ such that for every $\varepsilon > 0$ there is a left- neighbourhood U_ε of w_0 such that $\rho(\lambda(f_w - f_v)) < \varepsilon$ for every $w, v \in U_\varepsilon \cap W$. We say that the family $(f_w)_{w\in W}$ with $f_w \in \mathcal{X}_\rho$ for all $w \in W$ is $\|\cdot\|_\rho$-*Cauchy* (resp. $|||\cdot|||_\rho$-Cauchy), or satisfies the $\|\cdot\|_\rho$-*Cauchy condition with respect to the norm* $\|\cdot\|_\rho$ (resp. F-norm $|||\cdot|||_\rho$), if for every $\varepsilon > 0$ there is a left-neighbourhood U_ε of w_0 such that $\|f_w - f_v\|_\rho < \varepsilon$ (resp. $|||f_w - f_v|||_\rho < \varepsilon$) for every $v, w \in U_\varepsilon \cap W$. It is easily observed (compare with the proof of Theorem 1.2) that $(f_w)_{w\in W}$ is $\|\cdot\|_\rho$-Cauchy (resp. $|||\cdot|||_\rho$-Cauchy) if and only if for every $\lambda > 0$ and $\varepsilon > 0$ there exists a left-neighbourhood U_ε of w_0 such that $\rho(\lambda(f_w - f_v)) < \varepsilon$ for all $v, w \in U_\varepsilon \cap W$.

We may treat a family $\mathbb{T} = (T_w)_{w\in W}$ of convolution type operators (3.7) as a method of summability saying that a family $(f_w)_{w\in W}$ of functions $f_w \in \operatorname{Dom}\mathbb{T}$ is $(\mathbb{T}, \rho)$-*summable to a function* $f \in L^0_\rho(\Omega)$, if $T_w f_w \xrightarrow{w} f$ in the sense of the ρ-convergence, i.e. $\rho(\lambda(T_w f_w - f)) \xrightarrow{w} 0$ for sufficiently small $\lambda > 0$. A summability method generated by the family $\mathbb{T}$ of operators will also be called the $\mathbb{T}$-*method*. We shall say that the $\mathbb{T}$-method of summability is *conservative from* $L^0_\eta(\Omega)$ *to* $L^0_\rho(\Omega)$, or $\mathbb{T}$ is (η, ρ)-*conservative*, if for every family $(f_w)_{w\in W}$ such that $f_w \in L^0_\eta(\Omega) \cap \operatorname{Dom}\mathbb{T}$ for $w \in W$, η-convergent to a function $f \in L^0_\eta(\Omega)$, there exists a function $g \in L^0_\rho(\Omega)$ such that the family $(T_w f_w)_{w\in W}$ is ρ-convergent to g. If we have always $g = f$, we call the $\mathbb{T}$-method to be *regular* or *permanent from* $L^0_\eta(\Omega)$ *to* $L^0_\rho(\Omega)$. We say that a $\mathbb{T}$-method is *Cauchy conservative from* $L^0_\eta(\Omega)$ *to* $L^0_\rho(\Omega)$, or is (η, ρ)-*Cauchy conservative*, if for every family $(f_w)_{w\in W}$ such that $f_w \in L^0_\eta(\Omega) \cap \operatorname{Dom}\mathbb{T}$ for $w \in W$ which is η-Cauchy, there holds $T_w f_w \in L^0_\rho(\Omega)$ for $w \in W$ and the family $(T_w f_w)_{w\in W}$ is ρ-Cauchy. Instead of the pair of modulars (η, ρ) one may take norms (or F-norms) generated by these modulars, which leads to the notions of $(\|\cdot\|_\eta, \|\cdot\|_\rho)$-conservativeness and $(\|\cdot\|_\eta, \|\cdot\|_\rho)$-Cauchy conservativeness (or $(|||\cdot|||_\eta, |||\cdot|||_\rho)-$ conservativeness and $(|||\cdot|||_\eta, |||\cdot|||_\rho)$-Cauchy conservativeness) of a $\mathbb{T}$-method.

It is obvious that an (η, ρ)-conservative $\mathbb{T}$-method is (η, ρ)-Cauchy conservative, as well as that a $(\|\cdot\|_\eta, \|\cdot\|_\rho)$-conservative $\mathbb{T}$-method is $(\|\cdot\|_\eta, \|\cdot\|_\rho)$-Cauchy conservative and a $(|||\cdot|||_\eta, |||\cdot|||_\rho)$-conservative $\mathbb{T}$-method is $(|||\cdot|||_\eta, |||\cdot|||_\rho)$-Cauchy conservative. In order to obtain a converse result we need the space $L^0_\rho(\Omega)$ to be ρ-complete or $\|\cdot\|_\rho$-complete ($|||\cdot|||_\rho$-complete). We have to define what ρ-completeness of $L^0_\rho(\Omega)$ means. A sequence (f_n) of functions $f_n \in \mathcal{X}_\rho$, $n = 1, 2, \dots$, is called (ρ, λ)-*Cauchy with a constant* $\lambda > 0$, if for every $\varepsilon > 0$ there is an index N such that $\rho(\lambda(f_n - f_m)) < \varepsilon$ for all $m, n > N$. The modular space $\mathcal{X}_\rho$ is called ρ-*complete*, if for every $\lambda_1 > 0$ there exists a number $\lambda_2 > 0$ such that every sequence (f_n), $f_n \in \mathcal{X}_\rho$ for $n = 1, 2, \dots$, which is (ρ, λ_1)-Cauchy with λ_1, is ρ-convergent with λ_2 to an $f \in \mathcal{X}_\rho$, i.e. $\rho(\lambda_2(f_n - f)) \to 0$ as $n \to +\infty$. There holds the following

Theorem 5.2. (a) *If the modular space* $L^0_\rho(\Omega)$ *is* ρ*-complete, then every* ρ*-Cauchy family* $(f_w)_{w\in W}$ *is* ρ*-convergent to a function* $f \in L^0_\rho(\Omega)$.

(b) *If the modular space* $L^0_\rho(\Omega)$ *is* $\|\cdot\|_\rho$*-complete (resp.* $|||\cdot|||_\rho$*-complete), then* *every $\|\cdot\|_\rho$-Cauchy (resp. $|||\cdot|||_\rho$-Cauchy) family $(f_w)_{w\in W}$ is $\|\cdot\|_\rho$-convergent (resp. $|||\cdot|||_\rho$-convergent) to a function $f \in L^0_\rho(\Omega)$.*

Proof. We limit ourselves to the proof of (a). Let $(f_w)_{w\in W}$ be ρ-Cauchy, i.e. there exists a number $\lambda_1 > 0$ such that for every $\varepsilon > 0$ there is a left-neighbourhood U_ε of w_0 with the property that $\rho(\lambda_1(f_w - f)) < \varepsilon$ for all $w \in U_\varepsilon \cap W$. Let (w_n) be a sequence of elements of W such that $w_n \to w_0^-$ as $n \to +\infty$. Then $\rho(\lambda_1(f_{w_n} - f_{w_m})) < \varepsilon$ if m, n are so large that $w_n, w_m \in U_\varepsilon$. Hence the sequence (f_{w_n}) is (ρ, λ_1)-Cauchy with the constant λ_1. Since $L^0_\rho(\Omega)$ is ρ-complete, it follows that there exists an $f \in L^0_\rho(\Omega)$ and a number $\lambda_2 > 0$ such that $\rho(\lambda_2(f_{w_n} - f)) \to 0$ as $n \to +\infty$. Let us observe that the function f is independent of the sequence (w_n). Indeed should f and g be two limit functions relative to sequences (w_n) and (v_n), then

$$\rho\left(\frac{1}{3}\lambda_2(f-g)\right) \le \rho(\lambda_2(f - f_{w_n})) + \rho(\lambda_2(f_{w_n} - f_{v_n})) + \rho(\lambda_2(f_{v_n} - g)).$$

Since $(f_w)_{w\in W}$ is ρ-Cauchy, the right-hand side of the last inequality tends to 0 as $n \to +\infty$. Thus $f = g$. Since the sequence (w_n) was arbitrary, we conclude that $\rho(\lambda_2(f_w - f)) \xrightarrow{w} 0$. □

The following corollary may immediately be deduced from Theorem 5.2.

Corollary 5.1. (a) *If $L^0_\rho(\Omega)$ is ρ-complete, then any (η, ρ)-Cauchy conservative $\mathbb{T}$-method is (η, ρ)-conservative.*

(b) *If $L^0_\rho(\Omega)$ is $\|\cdot\|_\rho$-complete (or $|||\cdot|||_\rho$-complete), then any $(\|\cdot\|_\eta, \|\cdot\|_\rho)$-Cauchy conservative $((|||\cdot|||_\eta, |||\cdot|||_\rho)$-Cauchy conservative) $\mathbb{T}$-method is $(\|\cdot\|_\eta, \|\cdot\|_\rho)$-conservative $((|||\cdot|||_\eta, |||\cdot|||_\rho)$-conservative).*

In various circumstances it is advisable to consider a weaker notion than that of (η, ρ)-conservativeness of a $\mathbb{T}$-method of summability, considering not all families $(f_w)_{w\in W}$, but only uniformly bounded families. We say that a family $(f_w)_{w\in W}$ of functions $f_w \in L^0_\eta(\Omega)$ for $w \in W$ is uniformly bounded, if $f_w \in L^\infty(\Omega)$ for $w \in W$ and $\sup_{w\in W} \|f_w\|_\infty < +\infty$. A $\mathbb{T}$-method will be called a *boundedly (η, ρ)-conservative method* (resp. *boundedly (η, ρ)-Cauchy conservative method*), if for every uniformly bounded family $(f_w)_{w\in W}$, $f_w \in L^0_\eta(\Omega) \cap \mathbb{T}$ for $w \in W$, which is η-convergent to a function $f \in L^0_\eta(\Omega)$ (resp. which is η-Cauchy) there holds $T_w f_w \in L^0_\rho(\Omega)$ for $w \in W$ and the family $(T_w f_w)_{w\in W}$ is ρ-convergent to some $g \in L^0_\rho(\Omega)$ (resp. is ρ-Cauchy). Analogously, one may define that the $\mathbb{T}$-method is boundedly $(\|\cdot\|_\eta, \|\cdot\|_\rho)$-conservative or boundedly $(\|\cdot\|_\eta, \|\cdot\|_\rho)$-Cauchy conservative; the same may be done in the case of the F-norms $|||\cdot|||_\eta$ and $|||\cdot|||_\rho$.

In order to formulate a theorem on (η, ρ)-conservativeness and bounded (η, ρ)-conservativeness of a $\mathbb{T}$-method of summability we shall need still some special properties of the kernel $\mathbb{K} = (K_w)_{w\in W}$. We say that the Carathéodory kernel $\mathbb{K} = (K_w)_{w\in W}$

consists of *near kernel functions*, if for every $\delta > 0$ there exist a $w_\delta \in W$, $U_0 \subset \mathcal{U}_0$ such that for every $U \in \mathcal{U}_0$, $U \subset U_0$ and all $v, w \in [w_\delta, w_0[\cap W$, $u \in L^0(\Omega)$ there holds the inequality

$$\left| \int_U [K_w(t, u(t)) - K_v(t, u(t))]\, d\mu(t) \right| < \delta. \tag{5.2}$$

We say that the Carathéodory kernel $\mathbb{K} = (K_w)_{w \in W}$ consists of *almost near kernel functions*, if for every $\delta > 0$ and $\gamma > 0$, $\gamma < +\infty$, there exist a $w_{\delta,\gamma} \in W$ and $U_0 \in \mathcal{U}_0$, such that for every $U \in \mathcal{U}_0$, $U \subset U_0$ and all $v, w \in [w_{\delta,\gamma}, w_0[\cap W$, $|u(t)| \leq \gamma$ for μ-a.e. $t \in \Omega$ there holds the inequality (5.2). Obviously, if a kernel $\mathbb{K}$ consists of near kernel functions, then it consists of almost near kernel functions.

Theorem 5.3. *Let $\{\Omega, \mathcal{U}, \Sigma, \mu\}$ be a correctly filtered system with respect to the operation $+$. Let ρ be a monotone, J-quasiconvex modular on $L^0(\Omega)$ such that $\chi_\Omega \in L^0_\rho(\Omega)$. Let η be a modular on $L^0(\Omega)$, bounded with respect to the operation $+$ with a constant $C \geq 1$ and a function $\ell \in L^\infty(\Omega)$ such that $\ell(t) \xrightarrow{\mathcal{U}} 0$. Let $\mathbb{K} = (K_w)_{w \in W}$ be an $(\mathbb{L}, \psi)$-Lipschitz Carathéodory kernel, where $\psi \in \Psi$, $\mathbb{L} = (L_w)_{w \in W}$, $S = \sup_{w \in W} \|L_w\|_1 < +\infty$ and $p_w(t) = L_w(t)/\|L_w\|_1$ for $t \in \Omega$ satisfy the condition*

$$\int_{\Omega \setminus U} p_w(t)\, d\mu(t) \xrightarrow{w} 0 \quad \text{for every } U \in \mathcal{U}_0. \tag{5.3}$$

Suppose that $\{\rho, \psi, \eta\}$ is a properly directed triple. Let $\mathbb{T} = (T_w)_{w \in W}$, where T_w are defined by (3.7). *Then*

(a) *if $\mathbb{K}$ consists of near kernel functions, then the $\mathbb{T}$-method is (η, ρ)-Cauchy conservative;*

(b) *if $\mathbb{K}$ consists of almost near kernel functions, then the $\mathbb{T}$-method is boundedly (η, ρ)-Cauchy conservative.*

Proof. Let $(f_w)_{w \in W}$ be η-Cauchy and let $C \geq 1$ and $\lambda > 0$ be arbitrary. We have for arbitrary $w, w_1 \in W$ the inequality

$$\begin{aligned} \eta(2\lambda C f_w) &= \eta[2\lambda C(f_w - f_{w_1}) + 2\lambda C f_{w_1}] \\ &\leq \eta(4\lambda C(f_w - f_{w_1})) + \eta(4\lambda C f_{w_1}). \end{aligned}$$

As $(f_w)_{w \in W}$ is η-Cauchy, there exist $\lambda_0 > 0$ and $w_2 \in W$ with $\eta(4\lambda_0 C(f_w - f_{w_1})) < 1/2$ for $w, w_1 \in [w_2, w_0[\cap W$. Since $f_{w_1} \in L^0_\eta(\Omega)$, we may choose λ_0 so small that $\eta(4\lambda_0 C f_{w_1}) < 1/2$. Hence we have

$$\eta(2\lambda C f_w) < 1 \tag{5.4}$$

for $0 < \lambda \leq \lambda_0$ and $w \in [w_2, w_0[\cap W$. Now, we apply inequality (5.1), replacing $T_1, T_2, K_1, K_2, p_1, p_2, f, g$ by $T_w, T_v, K_w, K_v, p_w, p_v, f_w, f_v$, respectively. Applying (5.4), we obtain for $0 < \lambda \leq \lambda_0$ and $w, v \in [w_2, w_0[\cap W$ the inequality

$$\begin{aligned}
&\rho[\alpha(T_w f_w - T_v f_v)] \\
&\quad \leq \rho\left[6\alpha\left|\int_U (K_w(t, f_w(t+\cdot)) - K_v(t, f_w(t+\cdot)))\, d\mu(t)\right|\right] \\
&\qquad + M(1 + \|\ell\|_\infty)\left(\int_{\Omega\setminus U} p_w(t)\, d\mu(t) + \int_{\Omega\setminus U} p_v(t)\, d\mu(t)\right) \qquad (5.5)\\
&\qquad + M[\eta(C\lambda(f_w - f_v)) + \|\ell\chi_U\|_\infty] \\
&\qquad + 2M(1 + \|\ell\|_\infty)\int_{\Omega\setminus U} p_v(t)\, d\mu(t)
\end{aligned}$$

for every $U \in \mathcal{U}_0$.

Put $\gamma = \sup_{w\in W} \|f_w\|_\infty$; obviously, $\gamma < +\infty$ if and only if $(f_w)_{w\in W}$ is uniformly bounded. Let us take an arbitrary $\varepsilon > 0$. Since $\chi_\Omega \in L^0_\rho(\Omega)$, there exists a $\delta > 0$ such that

$$\rho(6\alpha\delta\chi_\Omega) < \varepsilon/7.$$

By assumptions (a) and (b) there exist $w_3 \in W$, $w_3 \geq w_2$ and $U_0 \in \mathcal{U}_0$ such that for arbitrary $U \in \mathcal{U}_0$, $U \subset U_0$, $v, w \in [w_3, w_0[\cap W$ and $u \in L^0(\Omega)$ such that $|u(t)| \leq \gamma$ μ-a.e. $t \in \Omega$, there holds the inequality (5.2). Here, $\gamma = +\infty$ in the case (a) and $\gamma < +\infty$, γ fixed, in the case (b). Thus we have

$$\left|\int_U (K_w(t, f_w(t+s)) - K_v(t, f_w(t+s)))\, d\mu(t)\right| < \delta$$

for $U \in \mathcal{U}_0$, $U \subset U_0$, $v, w \in [w_3, w_0[\cap W$ and all $s \in \Omega$. Hence we obtain, by monotonocity of ρ, the inequality

$$\rho\left[6\alpha\left|\int_U (K_w(t, f_w(t+\cdot)) - K_v(t, f_w(t+\cdot)))\, d\mu(t)\right|\right] \leq \rho(6\alpha\delta\chi_\Omega) < \varepsilon/7$$

for $U \in \mathcal{U}_0$, $U \subset U_0$, $v, w \in [w_3, w_0[\cap W$. Since $\ell(t) \xrightarrow{\mathcal{U}} 0$, we may suppose $U \in \mathcal{U}_0$ be so small that

$$\|\ell\chi_U\|_\infty < \frac{\varepsilon}{7M}.$$

We fix a set U satisfying the above inequalities. Since

$$\int_{\Omega\setminus U} p_w(t)\, d\mu(t) \xrightarrow{w} 0,$$

there exists a $w_4 \in W$, $w_4 \geq w_3$ such that

$$M(1 + \|\ell\|_\infty)\int_{\Omega\setminus U} p_w(t)\, d\mu(t) < \frac{\varepsilon}{7}, \quad M(1 + \|\ell\|_\infty)\int_{\Omega\setminus U} p_v(t)\, d\mu(t) < \frac{\varepsilon}{7}$$

for $w \in [w_4, w_0[\cap W$. Thus, the inequality (5.5) leads to an estimation

$$\rho[\alpha(T_w f_w - T_v f_v)] \leq \frac{6}{7}\varepsilon + M\eta(C\lambda(f_w - f_v))$$

for $v, w \in [w_4, w_0[\cap W$. Since $(f_w)_{w\in W}$ is η-Cauchy, there is a $w_5 \in W$, $w_5 \geq w_4$ such that $M\eta(C\lambda(f_w - f_v)) < \varepsilon/7$ for $v, w \in [w_5, w_0[\cap W$. Thus,

$$\rho[\alpha(T_w f_w - T_v f_v)] < \varepsilon$$

for $v, w \in [w_5, w_0[\cap W$. Consequently, the family $(T_w f_w)_{w\in W}$ is ρ-Cauchy. □

Let us remark that version (a) of Theorem 5.3 is less interesting in the case when the kernel functions are linear (Example 3.1), i.e., $K_w(t, u) = \widetilde{K}(t)u$. In this case inequality (5.2) becomes

$$\left|\int_U (\widetilde{K}_w(t) - \widetilde{K}_v(t))u(t)\, d\mu(t)\right| < \delta.$$

Taking $u(t) = u \in \mathbb{R}$, an arbitrary constant function, this is true only if $\int_U \widetilde{K}_w(t)\, d\mu(t) = \int_U \widetilde{K}_v(t)\, d\mu(t)$ for a $U \in \mathcal{U}_0$ and $w, v \in W$, w, v sufficiently near w_0.

5.3 Regularity of methods with respect to different modulars

Now we shall indicate a possibility to investigate the (η, ρ)-regularity of a $\mathbb{T}$-method directly, without applications of the Cauchy condition. The (η, ρ)-regularity of a $\mathbb{T}$-method means that if $(f_w)_{w\in W}$ with $f_w \in L^0_\eta(\Omega) \cap \operatorname{Dom} \mathbb{T}$ is η-convergent to an $f \in L^0_\eta(\Omega)$, then $(T_w f_w)_{w\in W}$ is ρ-convergent to f. This could be obtained applying the inequality

$$\rho[\alpha(T_w f_w - f)] \leq \rho[2\alpha(T_w f_w - f_w)] + \rho[2\alpha(f_w - f)],$$

if we only know that η-convergence $f_w \overset{w}{\to} f$ implies ρ-convergence $f_w \overset{w}{\to} f$; the first term on the right-hand side of the last inequality could be estimated by means of an inequality of the form (3.4).

In order to solve the problem, when η-convergence $f_w \overset{w}{\to} f$ implies ρ-convergence $f_w \overset{w}{\to} f$ we introduce the following relation. Let η and ρ be two modulars on $L^0(\Omega)$. We shall say that ρ is *weaker than* η, if there are positive constants α, β, γ and c such that for every $f \in L^0(\Omega)$ there holds the inequality

$$\rho(\alpha f) \leq \beta\eta(\gamma f) + c.$$

Moreover we will say that ρ is *strictly weaker than* η, if the previous relation holds with $c = 0$.

Example 5.1. Let us take

$$\rho(f) = \int_\Omega \varphi(|f(t)|)\, d\mu(t) \quad \text{and} \quad \eta(f) = \int_\Omega \psi(|f(t)|)\, d\mu(t),$$

where φ, ψ are φ-functions and $\mu(\Omega) < +\infty$ (Example 1.5 (b)). Let us suppose that there exist positive constants a, b, a_0, b_0 and u_0 such that

$$\varphi(au) \leq b\psi(a_0 u) + b_0$$

for every $u \geq u_0$. Let $f \in L^0(\Omega)$ and let $A = \{t \in \Omega : |f(t)| \geq u_0\}$, $B = \Omega \setminus A$. Then we have

$$\begin{aligned} \rho(af) &\leq \int_A \varphi(a|f(t)|)\, d\mu(t) + \int_B \varphi(au_0)\, d\mu(t) \\ &\leq \int_A b\psi(a_0|f(t)|)\, d\mu(t) + \int_A b_0\, d\mu(t) + \varphi(au_0)\mu(B) \\ &\leq b\eta(a_0 f) + (b_0 + \varphi(au_0))\mu(\Omega). \end{aligned}$$

This means that ρ is weaker than η. If $b_0 = 0$ and $u_0 = 0$ the modular ρ is clearly strictly weaker than η. In case when $\varphi(u) = u^p$ and $\psi(u) = u^q$ for $u \geq 0$ the above condition for φ and ψ means that $0 < p \leq q < +\infty$.

Theorem 5.4. *Let ρ and η be modulars on $L^0(\Omega)$ such that ρ is quasiconvex and weaker than η. Then $L^0_\eta(\Omega) \subset L^0_\rho(\Omega)$. Moreover if ρ is strictly weaker than η, the embedding is continuous both in the sense of modular convergence and in the sense of norm convergence in the spaces $L^0_\eta(\Omega)$ and $L^0_\rho(\Omega)$.*

Proof. Let $f \in L^0(\Omega)$ and $\lambda_0 > 0$ be such that $\eta(\lambda_0 f) < +\infty$. Let $\lambda > 0$ be such that $\lambda\gamma < \lambda_0$. Since ρ is weaker than η, we have

$$\rho(\alpha\lambda f) \leq \beta\eta(\gamma\lambda f) + c \leq \beta\eta(\lambda_0 f) + c < +\infty.$$

Since ρ is quasiconvex, with a constant $M \geq 1$, we finally deduce that $f \in L^0_\rho(\Omega)$.

Now let us assume that ρ is strictly weaker that η, i.e. $c = 0$. Let $g_n \in L^0_\eta(\Omega)$ for $n = 1, 2, \ldots$ be a sequence such that $g_n \to 0$ in the sense of η-convergence. Let $\lambda > 0$ be such that

$$\lim_{n \to +\infty} \eta(\gamma\lambda g_n) = 0.$$

Since ρ is strictly weaker than η we have

$$\rho(\alpha\lambda g_n) \leq \beta\eta(\gamma\lambda g_n),$$

and so the sequence $g_n \to 0$ in the sense of ρ-convergence.

The proof in the case of norm convergence follows the same lines. □

Arguing as in the proof of Theorem 5.4 one immediately obtains the following result.

Theorem 5.5. *Let ρ and η be modulars on $L^0(\Omega)$ such that ρ is quasiconvex and strictly weaker than η. Let $(f_w)_{w\in W}$ be a family of functions $f_w \in L^0_\eta(\Omega)$, $w \in W$.*

(a) *If $f_w \overset{w}{\to} f$ in the sense of η-convergence, then $f_w \overset{w}{\to} f$ in the sense of ρ-convergence, too.*

(b) *If $f_w \overset{w}{\to} f$ in the sense of $\|\cdot\|_\eta$-convergence (or $|||\cdot|||_\eta$-convergence), then $f_w \overset{w}{\to} f$ in the sense of $\|\cdot\|_\rho$-convergence (or $|||\cdot|||_\rho$-convergence).*

We may now prove the following theorem on regularity of a $\mathbb{T}$-method.

Theorem 5.6. *Let $\{\Omega, \mathcal{U}, \Sigma, \mu\}$ be a correctly filtered system with respect to the operation $+$. Let ρ be a monotone, J-quasiconvex with constant $M \geq 1$ modular on $L^0(\Omega)$ and let η be an absolutely finite, absolutely continuous and bounded modular on $L^0(\Omega)$ such that ρ is strictly weaker than η. Let $\mathbb{K} = (K_w)_{w\in W}$ be an $(\mathbb{L}, \psi)$-Lipschitz, strongly singular Carathéodory kernel such that $\{\rho, \psi, \eta\}$ is a properly directed triple. Then the $\mathbb{T}$-method defined by the family $\mathbb{T} = (T_w)_{w\in W}$ of operators given by* (3.7) *is (η, ρ)-regular.*

Proof. By Theorem 3.2, inequality (3.4) and Lemma 3.1 (a) we have

$$\begin{aligned}\rho[\alpha(T_w f_w - f_w)] \leq M\omega_\eta(\lambda f_w, U) + M[2\eta(2\lambda C f_w) \\ + \|\ell\|_\infty] \int_{\Omega\setminus U} p_w(t)\, d\mu(t) + \rho(2\alpha r_w^{(0)} f_w), \quad w \in W,\end{aligned} \tag{5.6}$$

for $U \in \Sigma$, $0 < \lambda < 1$, $0 < \alpha < C_\lambda(2MS)^{-1}$ and $f_w \in L^0_\eta(\Omega) \cap \operatorname{Dom}\mathbb{T}$, where $S = \sup_{w\in W} \|L_w\|_1 < +\infty$. First, we estimate $\omega_\eta(\lambda f_w, U)$. By the boundedness of η, we have for $t \in U$ and $w \in W$

$$\begin{aligned}\eta[\lambda(f_w(t+\cdot) - f_w(\cdot))] &\leq \eta[3\lambda(f_w(t+\cdot) - f(t+\cdot))] \\ &\quad + \eta[3\lambda(f(t+\cdot) - f(\cdot))] + \eta[3\lambda(f(\cdot) - f_w(\cdot))] \\ &\leq \eta[3\lambda C(f_w - f)] + \ell(t) + \eta[3\lambda(f(t+\cdot) - f(\cdot))] \\ &\quad + \eta[3\lambda(f_w - f)] \\ &\leq 2\eta[3\lambda C(f_w - f)] + \|\ell\chi_U\|_\infty + \omega_\eta(3\lambda f, U),\end{aligned}$$

for $w \in W$. Hence

$$\omega_\eta(\lambda f_w, U) \leq 2\eta[3\lambda C(f_w - f)] + \|\ell\chi_U\|_\infty + \omega_\eta(3\lambda f, U).$$

Inserting the above inequality in (5.6), we obtain the inequality

$$\begin{aligned}\rho[\alpha(T_w f_w - f_w)] &\leq 2M\eta[3\lambda C(f_w - f)] + M\|\ell\chi_U\|_\infty \\ &\quad + M\omega_\eta(3\lambda f, U) + M[2\eta(2\lambda C f_w) \\ &\quad + \|\ell\|_\infty] \int_{\Omega\setminus U} p_w(t)\, d\mu(t) + \rho(2\alpha r_w^{(0)} f_w).\end{aligned} \tag{5.7}$$

Let $\varepsilon > 0$ be given. Since η is bounded, we have $\|\ell\chi_U\|_\infty \xrightarrow{\mathcal{U}} 0$. Moreover, by Theorem 2.4, we have $\omega_\eta(2\lambda f, U) \xrightarrow{\mathcal{U}} 0$ for sufficiently small $\lambda > 0$. Hence there exist a set $U \in \mathcal{U}_0$ and $\lambda_1 \in]0, 1[$ such that

$$M\|\ell\chi_U\|_\infty < \varepsilon/6 \quad \text{and} \quad M\omega_\eta(3\lambda f, U) < \varepsilon/6$$

for $0 < \lambda \leq \lambda_1$. Since $f_w \overset{w}{\to} f$ in the sense of η-convergence, there exist $w_1 \in W$ and $0 < \lambda_2 \leq \lambda_1$ such that

$$2M\eta[3\lambda C(f_w - f)] < \varepsilon/6$$

for $w \in [w_1, w_0[\cap W$ and $0 < \lambda \leq \lambda_2$. Moreover, arguing as in the proof of Theorem 5.3 we may easily show that taking λ_2 sufficiently small and w_1 sufficiently near to w_0 we have

$$\eta(2\lambda C f_w) < 1$$

for $\lambda \in]0, \lambda_2[$ and $w \in [w_1, w_0[\cap W$. Hence, from (5.7) we obtain the inequality

$$\rho[\alpha(T_w f_w - f_w)] \leq \frac{\varepsilon}{2} + M(2 + \|\ell\|_\infty) \int_{\Omega\setminus U} p_w(t)\, d\mu(t) + \rho(2\alpha r_w^{(0)} f_w) \quad (5.8)$$

for $w \in [w_1, w_0[\cap W$, where $U \in \mathcal{U}_0$ is the set we have fixed above, $0 < \lambda < \lambda_2$ and $0 < \alpha < C_\lambda(2MS)^{-1}$. Since $\mathbb{K}$ is singular, we may take a $w_2 \in W$ such that $w_2 \in [w_1, w_0[$ and

$$M[2 + \|\ell\|_\infty] \int_{\Omega\setminus U} p_w(t)\, d\mu(t) < \varepsilon/6$$

for $w \in [w_2, w_0[\cap W$, where U is the above fixed set in $\mathcal{U}_0$. Hence, by (5.8), we obtain

$$\rho[\alpha(T_w f_w - f_w)] \leq \frac{2}{3}\varepsilon + \rho(2\alpha r_w^{(0)} f_w)$$

for $\alpha > 0$ as above, say $0 < \alpha \leq \alpha_0$ and $w \in [w_2, w_0[\cap W$. Thus, if $0 < \alpha \leq \alpha_0$ and $w_2 \leq w < w_0$, $w \in W$, we have

$$\rho[\alpha(T_w f_w - f_w)] \leq \frac{2}{3}\varepsilon + \rho[4\alpha r_w^{(0)}(f_w - f)] + \rho(4\alpha r_w^{(0)} f). \quad (5.9)$$

Since ρ is quasiconvex and strictly weaker than η so, by Theorem 5.4, $L^0_\eta(\Omega) \subset L^0_\rho(\Omega)$ and this embedding is continuous in the sense of modular convergence. Consequently, since $f_w \in L^0_\eta(\Omega)$ for $w \in W$, $f \in L^0_\eta(\Omega)$ and $f_w \overset{w}{\to} f$ in the sense of η-convergence, we obtain $f \in L^0_\rho(\Omega)$ and $f_w \overset{w}{\to} f$ in the sense of ρ-convergence. Thus, there exist $\beta > 0$ and $w_3 \in W$ with $w_3 \geq w_2$, such that $\rho(\beta f) < \varepsilon/6$ and $\rho[\beta(f_w - f)] < \varepsilon/6$ for $w \in [w_3, w_0[\cap W$. Since $\mathbb{K}$ is strongly singular, there holds $r_w^{(0)} \overset{w}{\to} 0$. Hence there is a $w_4 \in W$, $w_4 \geq w_3$ such that $4\alpha_0 r_w^{(0)} < \beta$ for $w \in [w_4, w_0[\cap W$. Consequently,

$$\rho(4\alpha_0 r_w^{(0)} f) < \varepsilon/6 \quad \text{and} \quad \rho[4\alpha_0 r_w^{(0)}(f_w - f)] < \varepsilon/6$$

for $w \in [w_4, w_0[\cap W$. Hence we obtain, applying the inequality (5.9),

$$\rho[\alpha(T_w f_w - f)] < \varepsilon$$

for $0 < \alpha \leq \alpha_0$, $w \in [w_4, w_0[\cap W$. Thus, choosing for example $\gamma < \alpha$, we may write

$$\rho[\frac{\gamma}{2}(T_w f_w - f)] \leq \rho[\gamma(T_w f_w - f_w)] + \rho[\gamma(f_w - f)]$$

and thus $T_w f_w \overset{w}{\rightarrow} f$ in the sense of ρ-convergence □

Let us still remark that similarly as in Theorem 3.3, the assumption of strong singularity of the kernel $\mathbb{K}$ may be replaced by a weaker one of singularity; however, the modular ρ must be finite and absolutely continuous. We are not going into details of the formulation of the theorem in this case. Let us also remark that Theorems 5.3 and 5.6 are not referred to summability methods $\mathbb{T} = (T_w)_{w \in W}$ with $T_w = T$ for $w \in W$, since in this case the condition (5.3) is never satisfied.

Example 5.2. Let us take $\Omega = \mathbb{Z}$ the set of all integers, Σ the σ-algebra of all subsets of $\mathbb{Z}$, μ the counting measure on $\mathbb{Z}$. Then we have for $f(j) = a_j$, $j \in \mathbb{Z}$,

$$\int_\Omega f(t)\, d\mu(t) = \sum_{j \in \mathbb{Z}} a_j.$$

Let $W = \mathbb{N}$ the set of all positive integers, $w_0 = +\infty$. We take a linear kernel $\mathbb{K} = (K_n)_{n=1}^{\infty}$, i.e. $K_n(j, u) = a_j^{(n)} u$ for $j \in \mathbb{Z}$, $n \in \mathbb{N}$. Then obviously $|K_n(j, u) - K_n(j, v)| \leq |a_j^{(n)}||u - v|$ for $u, v \in \mathbb{R}$, $j \in \mathbb{Z}$, $n \in \mathbb{N}$. This means that the kernel $\mathbb{K}$ is $(\mathbb{L}, \psi)$-Lipschitz with $\mathbb{L} = (L_n)_{n=1}^{\infty}$, where $L_n(j) = |a_j^{(n)}|$, and with $\psi(j, u) = u$ for $u \geq 0$. We have $\|L_n\|_1 = \sum_{j \in \mathbb{Z}} |a_j^{(n)}|$, whence the condition $L_n \in L^1(\Omega)$ for $n \in \mathbb{N}$ means that the series $\sum_{j \in \mathbb{Z}} a_j^{(n)}$ are absolutely convergent for $n \in \mathbb{N}$ and the condition $S = \sup_{n \in \mathbb{N}} \|L_n\|_1 < +\infty$ means that $\sup_n \sum_{j \in \mathbb{Z}} |a_j^{(n)}| < +\infty$. Moreover, we have $p_n(j) = \|L_n\|_1^{-1} L_n(j)$ for $j \in \mathbb{Z}$, $n \in \mathbb{N}$. Now, let $\mathcal{U}$ be the family of sets $U \subset \mathbb{Z}$ such that $\mathbb{Z} \setminus U$ is finite or empty. Taking $U_k = \{j \in \mathbb{Z} : |j| > k\}$ for $k = 1, 2, \ldots$ we see that $\mathcal{U}_0 = \{U_1, U_2, U_3, \ldots\}$ is a basis of the filter $\mathcal{U}$. We are going to interpret what singularity and strong singularity of the kernel $\mathbb{K}$ means. We have for $U = U_k \in \mathcal{U}_0$

$$\int_{\Omega \setminus U} p_w(t)\, d\mu(t) = \sum_{j=-k}^{k} \frac{|a_j^{(n)}|}{\|L_n\|_1} \quad \text{for } n \in \mathbb{N},$$

and so the condition (5.3) is equivalent to

$$\lim_{n \to +\infty} \sum_{j=-k}^{k} \frac{|a_j^{(n)}|}{\|L_n\|_1} = 0$$

for $k = 1, 2, \ldots$, and this is equivalent to the condition

$$\lim_{n\to+\infty} \frac{|a_j^{(n)}|}{\|L_n\|_1} = 0, \quad \text{for every } j \in \mathbb{Z}. \tag{5.10}$$

Moreover, we have for $l = 1, 2, \ldots$

$$r_n^{(l)} = \sup_{l^{-1}\le|u|\le l} \Big|\frac{1}{u}\sum_{j\in\mathbb{Z}} a_j^{(n)} u - 1\Big| = \Big|\sum_{j\in\mathbb{Z}} a_j^{(n)} - 1\Big| = r_n^{(0)}$$

for $n = 1, 2, \ldots$. Hence singularity of $\mathbb{K}$ is equivalent to strong singularity of $\mathbb{K}$ and it is given by the conditions (5.10) and

$$\lim_{n\to+\infty} \sum_{j\in\mathbb{Z}} a_j^{(n)} = 1. \tag{5.11}$$

Let us still examine the assumption that the kernel $\mathbb{K} = (K_n)_{n=1}^{\infty}$ consists of almost near kernel functions. Let $\delta > 0$ and $\gamma > 0$ be given and $U \subset U_k$; then the inequality (5.2) is equivalent to

$$\Big|\sum_{j\in U}(a_j^{(n)} - a_j^{(m)})\Big| < \frac{\delta}{\gamma}.$$

Thus, the assumption on $\mathbb{K}$ to consist of almost near kernel functions means that for every $\varepsilon > 0$ there exist $N \in \mathbb{N}$ and $k \in \mathbb{N}$ such that for every set $U \subset \mathbb{Z}$ with $|j| \ge k$ for all $j \in U$ and for any $m, n > N$ there holds the inequality

$$\Big|\sum_{j\in U} a_j^{(n)} - \sum_{j\in U} a_j^{(m)}\Big| < \varepsilon.$$

5.4 Bibliographical notes

Problems of summability by means of families $(T_w)_{w\in W}$ of convolution-type nonlinear integral operators were started in [159], and continued by B. Tomasz [190], [191]. Theorem 5.1 with the estimation (5.1) was proved as Theorem 6 in [147], Theorem 5.3 as Theorem 8 in [147]. The contents of Section 5.3 have not been published.

Chapter 6

Nonlinear integral operators in the space BV_φ

6.1 Preliminaries

In this chapter we will discuss some properties concerning nonlinear integral operators of convolution type applied to functions belonging to the space of functions with bounded φ-variation (see Example 1.5(f)) on an interval $I \subset \mathbb{R}$.

For the sake of simplicity we will consider the (unbounded) interval $I =]0, +\infty[$, endowed with its Lebesgue measure. As in Example 1.5 (f), we denote by $\mathcal{X}$ the space of all real valued functions defined on I. However, we always assume that the functions in $\mathcal{X}$ are measurable. We recall here the definition of the φ-variation of f. Let φ be a φ-function (see Example 1.5 (b)). We define

$$V_\varphi(f) \equiv V_\varphi(f, I) = \sup_{\Pi} \sum_{j=1}^{n} \varphi(|f(s_j) - f(s_{j-1})|), \tag{6.1}$$

where the supremum runs over all finite increasing sequences $\{s_1, \dots, s_n\}$, with $s_j \in I$, $j = 1, \dots, n$.

Recall that the functional $f \longmapsto V(f)$ is not a modular functional on $\mathcal{X}$, because $V_\varphi(f) = 0$ only implies $f =$ constant. However, as remarked in Example 1.5(f), it is easy to show that the functional $\rho : \mathcal{X} \to \mathbb{R}_0^+$ defined by

$$\rho(f) = |f(a)| + V_\varphi(f), \tag{6.2}$$

for every $f \in \mathcal{X}$ and a fixed $a \in I$, is a modular functional. The corresponding modular space $\mathcal{X}_\rho$ is the space of functions with bounded φ-variation in I and it is usually denoted by $BV_\varphi(I)$.

Alternatively, it is possible to define the above modular space, by observing that two functions f, g such that $f - g =$ constant, have the same φ-variation. Thus, introducing in $\mathcal{X}$ the following equivalence relation

$$f \sim g \quad \text{iff} \quad f - g = \text{constant},$$

the functional $V_\varphi(f)$ is a modular on the quotient space $\mathcal{X}/\sim$. Using this approach, we identify functions which differ by an additive constant.

Let us remark firstly that functions $f \in BV_\varphi(I)$ are bounded and the limit $\lim_{t\to 0^+} f(t)$ exists and it is finite. We will prove the second assertion (boundedness being trivial).

Obviously, if the limit exists, then it is finite, otherwise the function f could not be bounded. Now, putting $L = \overline{\lim}_{t\to 0^+} f(t)$, $l = \underline{\lim}_{t\to 0^+} f(t)$, if the assertion is false, then $\tau = L - l > 0$. Then there are two decreasing sequences $\{t_i\}_{i\in\mathbb{N}}$ and $\{s_j\}_{j\in\mathbb{N}}$ in I, with $t_i \to 0$ and $s_j \to 0$, such that $f(s_j) \to L$ and $f(t_i) \to l$. We can choose the points t_i and s_j in such a way that $t_{i-1} < s_j < t_i$. In this way for sufficiently large i, j we have, for any $\lambda > 0$, $\varphi(\lambda|f(s_j) - f(t_i)|) > \varphi(\lambda(\tau/2))$. Taking a finite sequence of numbers $\{\xi_1, \dots \xi_N\}$ where $\{\xi_\nu\}$ are alternatively of type t_i and s_j for sufficiently large indices i, j, we obtain

$$\sum_{\nu=1}^{N-1} \varphi(\lambda|f(\xi_\nu) - f(\xi_{\nu+1})|) > (N-1)\varphi(\lambda(\tau/2)),$$

and so $f \notin BV_\varphi(I)$.

As a consequence, we can define the modular functional (6.2) using the formula

$$\rho(f) = |f(0+)| + V_\varphi(f).$$

6.2 Some estimates in BV_φ

One of the main problems concerning the modular space $BV_\varphi(I)$ is that the generating modular (6.2), (or the functional (6.1)), is not monotone. Moreover it is neither finite nor absolutely continuous. Thus the study of the properties of nonlinear integral operators in this setting is quite difficult, and a large part of the theory developed in the previous chapters, leads actually to various open problems.

In this section we will obtain some inequality related to the functional (6.1). As consequences we will obtain some embedding theorems.

At first we introduce the class of operators. Let $\widetilde{K} : I \times I \times \mathcal{X} \to \mathbb{R}$ be the functional defined by

$$\widetilde{K}(s, t, f) = K(t, f(st)),$$

for any $s, t \in I$ and $f \in \mathcal{X}$, where the function $K : I \times \mathbb{R} \to \mathbb{R}$ is a measurable function satisfying a (L, ψ)-Lipschitz condition of type

$$|K(t, u) - K(t, v)| \le L(t)\psi(t, |u - v|),$$

for every $t \in \mathbb{R}^+$, $u, v \in \mathbb{R}$, where $L \in L^1(\mathbb{R}^+)$ and the function ψ belongs to the class Ψ (see Section 3.1). We will assume *always* that ψ is a *concave* function with respect to the second variable. Note that we do not need the assumption that $\widetilde{K}(s, t, 0) = 0$. We will denote the class of such functions K by $\mathcal{K}$. We define our operator by means of the formula

$$(Tf)(s) = \int_{\mathbb{R}^+} K(t, f(st))\, dt, \tag{6.3}$$

where $f \in \operatorname{Dom} T$, the set of all functions $f \in \mathcal{X}$ such that Tf is well defined and $|(Tf)(s)| < +\infty$, for every $s \in \mathbb{R}^+$. In order to obtain some inequalities, that lead to embeddings theorems, we introduce an assumption similar to those used in Section 1.4. Given a φ-function $\varphi : \mathbb{R}_0^+ \to \mathbb{R}_0^+$, we will assume the following condition.

There are a φ-function $\gamma : \mathbb{R}_0^+ \to \mathbb{R}_0^+$ and a measurable function $v : \mathbb{R}^+ \to \mathbb{R}_0^+$ such that

$$\varphi(\psi(t,u)) \leq v(t)\gamma(u), \quad \textit{for every } u \in \mathbb{R}_0^+,\ t \in \mathbb{R}^+,\ \psi \in \Psi. \tag{6.4}$$

Note that this assumption does not imply that the modulars generated by the variations V_φ, V_γ and the function ψ determine a properly directed triple (see Section 1.4).

The following estimation relates the spaces $BV_\varphi(I)$ and $BV_\gamma(I)$; namely we establish that T maps $\operatorname{Dom} T \cap BV_\gamma(I)$ into $BV_\varphi(I)$.

Theorem 6.1. *Let $K \in \mathcal{K}$, φ a convex φ-function, $\psi \in \Psi$ and let γ be a φ-function satisfying condition* (6.4). *Let moreover a be a constant with $0 < a \leq 1/A_L$, where*

$$0 < A_L := \int_{\mathbb{R}^+} L(t)\,dt < +\infty,\ A_L^v = \int_{\mathbb{R}^+} L(t)v(t)\,dt < +\infty.$$

Then, if $f \in \operatorname{Dom} T$, there results

$$V_\varphi[a(Tf)] \leq A_L^{-1} A_L^v V_\gamma[aA_L f].$$

Proof. We may suppose $V_\gamma[aA_L f] < +\infty$. Let $D = \{s_i\}_{i=0,1,\dots N}$ be an increasing finite sequence in I. Fixed arbitrarily an index $i \in \{1, 2, \dots N\}$, we obtain, from the Lipschitz condition of K,

$$\begin{aligned} |(Tf)(s_i) - (Tf)(s_{i-1})| &\leq \int_{\mathbb{R}^+} |K(t, f(s_i t)) - K(t, f(s_{i-1}t))|\,dt \\ &\leq \int_{\mathbb{R}^+} L(t)\psi(t, |f(s_i t) - f(s_{i-1}t)|)\,dt. \end{aligned}$$

Now, by monotonicity of φ, Jensen's inequality and concavity of $\psi(t,\cdot)$, we obtain:

$$\begin{aligned} &\sum_{i=1}^N \varphi(a|(Tf)(s_i) - (Tf)(s_{i-1})|) \\ &\quad\leq \sum_{i=1}^N \varphi(a \int_{\mathbb{R}^+} L(t)\psi(t, |f(s_i t) - f(s_{i-1}t)|)\,dt) \\ &\quad\leq \frac{1}{A_L}\int_{\mathbb{R}^+} L(t)\sum_{i=1}^N \varphi(aA_L\psi(t, |f(s_i t) - f(s_{i-1}t)|))\,dt \leq \end{aligned}$$

$$\leq \frac{1}{A_L}\int_{\mathbb{R}^+} L(t)\sum_{i=1}^{N}\varphi(\psi(t, aA_L|f(s_it)-f(s_{i-1}t)|))\,dt$$

$$\leq \frac{1}{A_L}\int_{\mathbb{R}^+} L(t)v(t)\sum_{i=1}^{N}\gamma(aA_L|f(s_it)-f(s_{i-1}t)|)\,dt$$

$$\leq \frac{1}{A_L}\int_{\mathbb{R}^+} L(t)v(t)V_\gamma[aA_Lf]\,dt = A_L^{-1}A_L^{v}V_\gamma[aA_Lf];$$

the assertion follows by the arbitrariness of the set D. □

Remark 6.1. Let now consider the particular case when $\psi(t,u)=u$, for every $t\in\mathbb{R}^+$ and $u\in\mathbb{R}_0^+$; then the Lipschitz condition takes now the form

$$|K(t,u)-K(t,v)|\leq L(t)|u-v|$$

for every $t\in\mathbb{R}^+$, $u,v\in\mathbb{R}$, being $L:\mathbb{R}^+\to\mathbb{R}_0^+$ a summable function. This is the so-called "strongly-Lipschitz" condition. In this case condition (6.4) is satisfied as an equality with $\gamma(u)=\varphi(u)$ and $v(t)=1$ for every $t\in\mathbb{R}^+$. Hence, $A_L=A_L^v$ and so Theorem 6.1 gives the estimate: $V_\varphi[a(Tf)]\leq V_\varphi[aA_Lf]$ where a is a positive constant. So in this case Theorem 6.1 states that T maps $\operatorname{Dom}T\cap BV_\varphi(\mathbb{R}^+)$ in $BV_\varphi(\mathbb{R}^+)$. The same estimate is obtained in the particular case of a linear operator of the form

$$(Mf)(s)=\int_{\mathbb{R}^+}K(t)f(st)\,dt$$

when $0<A_L:=\int_{\mathbb{R}^+}K(t)\,dt<+\infty$. If $A_L\equiv 1$, as it happens for moment type kernels, i.e. $K_\lambda(t)=\lambda t^{\lambda-1}\chi_{]0,1[}(t)$, $t\in\mathbb{R}^+$, $\lambda>1$, and $a=1$, then Theorem 6.1 gives

$$V_\varphi[M_\lambda f]\leq V_\varphi[f] \tag{6.5}$$

where $(M_\lambda f)(s)=\int_{\mathbb{R}^+}K_\lambda(t)f(st)\,dt$.

Inequality (6.5) is the so-called *φ-variation-non augmenting (diminishing) property* for $M_\lambda f$.

6.3 A superposition theorem in BV_φ

In fractional calculus some generalized concepts of variation are used. In particular, by using some classical *linear* integral operators, like Riemann–Liouville fractional integrals, as example, a concept of *fractional variation* was introduced in [212]. It is defined simply by taking the variation of the transformed function under the integral transform considered. Thus, for example, denoting by $\mathcal{P}_\alpha f$ the Riemann–Liouville integral of f of order $\alpha\in]0,1[$, defined by

$$(\mathcal{P}_\alpha f)(s)=\frac{1}{\Gamma(1-\alpha)}\int_0^s\frac{f(t)}{(s-t)^\alpha}\,dt,\quad s\in\mathbb{R}^+,\ f\in L^1_{\mathrm{loc}}(\mathbb{R}^+),$$

the α-variation of f in I is defined by

$$V^\alpha(f) = V^\alpha(f, I) = V(\mathcal{P}_\alpha f).$$

The Riemann–Liouville fractional integral is a linear integral operator with kernel

$$H(s,t) = \frac{1}{\Gamma(1-\alpha)} \frac{1}{(s-t)^\alpha} \chi_{]0,s[}(t),$$

and the domain of this operator is $L^1_{\mathrm{loc}}(\mathbb{R}^+)$ (see [145]).

It is remarkable that this concept is used in order to describe some geometric properties (e.g. Hausdorff dimension) of some fractal sets (see e.g. [99]). For this concept of variation, some "variation diminishing" properties for integral operators were given (see e.g. [212]).

We can obtain a general approach to this problem, by introducing a concept of (U, φ)-variation, where U is a general nonlinear integral operator. So given U we can define

$$V^U_\varphi(f) = V_\varphi(Uf),$$

for any f belonging to suitable subspaces of $\mathfrak{X}$.

Thus it seems to be very natural to study some "variation diminishing" properties for these concepts of variation. This involves some estimate concerning the superposition of integral operators. Here we give some contribution to this problem.

We consider the following nonlinear integral operators

$$(Uf)(s) = \int_{\mathbb{R}^+} H(t, f(st))\, dt, \quad f \in \operatorname{Dom} U$$

and

$$(Tf)(s) = \int_{\mathbb{R}^+} K(t, f(st))\, dt, \quad f \in \operatorname{Dom} T,$$

being $H, K : \mathbb{R}^+ \times \mathbb{R} \to \mathbb{R}$ measurable functions with $H, K \in \mathcal{K}$, where H is (M, θ)-Lipschitz, i.e. there exist $\theta \in \Psi$, (concave with respect to the second variable), and a measurable function $M : \mathbb{R}^+ \to \mathbb{R}^+_0$ such that

$$|H(t,u) - H(t,v)| \le M(t)\theta(t, |u-v|),$$

for every $t \in \mathbb{R}^+$, $u, v \in \mathbb{R}$. Moreover we assume $Tf \in \operatorname{Dom} U$. In order to establish an estimate for the composition of the nonlinear integral operators, we will need of the following condition $(\star)$.

Let φ be a convex φ-function, $\psi, \theta \in \Psi$; there exist $\sigma, \lambda : \mathbb{R}^+_0 \to \mathbb{R}^+_0$, where σ is a convex φ-function and λ is a φ-function, and measurable functions $v_1, v_2 : \mathbb{R}^+ \to \mathbb{R}^+_0$, such that the following relations are satisfied

$$\varphi(\theta(t,u)) \le v_1(t)\sigma(u), \ \sigma(\psi(t,u)) \le v_2(t)\lambda(u), \quad u \in \mathbb{R}^+_0, t \in \mathbb{R}^+. \qquad (\star)$$

Example 6.1. A class of functions satisfying the conditions $(\star)$ is given for example by taking $\varphi(u) = e^{u^\alpha} - 1, \alpha \geq 1$ and $\theta \in \Psi$ of type $\theta(t, u) = \theta(u)$. So we put $\sigma(u) = \varphi(\theta(u)) = e^{\theta^\alpha(u)} - 1$, $v_i = 1, i = 1, 2$ for every $t \in \mathbb{R}^+$, and there results that φ is a convex φ-function and $\sigma(u)$ may be a convex (or not convex) φ-function, (this depends on the form of the function θ and on the exponent α).

Now we are ready to state a sufficient condition under which the operator $(U \circ T)$ maps $\operatorname{Dom} T \cap BV_\lambda(\mathbb{R}^+)$ in $BV_\varphi(\mathbb{R}^+)$.

Theorem 6.2. *Let $H, K \in \mathcal{K}$, with H (M, θ)-Lipschitz, $\theta \in \Psi$. Let φ be a convex φ-function, $\psi \in \Psi$ and let σ and λ be the functions of the condition $(\star)$. Furthermore let us suppose*

$$\text{i) } 0 < A_L := \int_{\mathbb{R}^+} L(t)\,dt < +\infty,\ A_L^{v_2} := \int_{\mathbb{R}^+} L(t)v_2(t)\,dt < +\infty,$$

$$\text{ii) } 0 < A_M := \int_{\mathbb{R}^+} M(t)\,dt < +\infty,\ A_M^{v_1} := \int_{\mathbb{R}^+} M(t)v_1(t)\,dt < +\infty,$$

and let a be a constant with $0 < a \leq \min\left\{\frac{1}{A_L A_M}, \frac{1}{A_M}\right\}$. Then, if $f \in \operatorname{Dom} T$, there results

$$V_\varphi[a(U \circ T)f] \leq \frac{A_M^{v_1}}{A_M} V_\sigma[aA_M(Tf)] \leq \frac{A_M^{v_1} A_L^{v_2}}{A_M A_L} V_\lambda[aA_M A_L f].$$

Proof. We may suppose $V_\lambda[aA_M A_L f] < +\infty$; let $D = \{s_i\}_{i=0,1,\ldots,N}$ be a finite increasing sequence in $\mathbb{R}^+$. Fixed arbitrarily an index $i \in \{1, 2, \ldots, N\}$, we have

$$\begin{aligned} |(U \circ T)f(s_i) - (U \circ T)f(s_{i-1})| &\leq \int_{\mathbb{R}^+} |H(t, (Tf)(s_i t)) - H(t, (Tf)(s_{i-1}t))|\,dt \\ &\leq \int_{\mathbb{R}^+} M(t)\theta(t, |(Tf)(s_i t) - (Tf)(s_{i-1}t)|)dt. \end{aligned}$$

Now, by monotonicity of φ, Jensen's inequality, concavity of $\theta(t, \cdot)$ and condition $(\star)$, we may write

$$\begin{aligned} &\sum_{i=1}^N \varphi(a|(U \circ T)f(s_i) - (U \circ T)f(s_{i-1})|) \\ &\quad \leq \sum_{i=1}^N \varphi(a \int_{\mathbb{R}^+} M(t)\theta(t, |(Tf)(s_i t) - (Tf)(s_{i-1}t)|)dt) \\ &\quad \leq \frac{1}{A_M} \int_{\mathbb{R}^+} M(t) \sum_{i=1}^N \varphi(aA_M \theta(t, |(Tf)(s_i t) - (Tf)(s_{i-1}t)|))\,dt \leq \end{aligned}$$

$$\le \frac{1}{A_M}\int_{\mathbb{R}^+} M(t)\sum_{i=1}^N \varphi(\theta(t, aA_M|(Tf)(s_i t) - (Tf)(s_{i-1}t)|))\,dt$$

$$\le \frac{1}{A_M}\int_{\mathbb{R}^+} M(t)v_1(t)\sum_{i=1}^N \sigma(aA_M|(Tf)(s_i t) - (Tf)(s_{i-1}t)|)\,dt$$

$$\le \frac{1}{A_M}\int_{\mathbb{R}^+} M(t)v_1(t)V_\sigma[aA_M(Tf)]\,dt = \frac{A_M^{v_1}}{A_M}V_\sigma[aA_M(Tf)].$$

Applying Theorem 6.1, there results

$$\frac{A_M^{v_1}}{A_M}V_\sigma[aA_M(Tf)] \le \frac{A_M^{v_1}}{A_M}\frac{A_L^{v_2}}{A_L}V_\lambda[aA_M A_L f],$$

and the assertion follows by the arbitrariness of the sequence D. □

Remark 6.2. In the particular case when $\theta(t,u) = \psi(t,u) = u$, for $u \in \mathbb{R}^+$, then in condition ($\star$) we may take $\sigma(u) = \lambda(u) = \varphi(u)$ and $v_i(t) = 1, i = 1, 2$ and hence we obtain

$$V_\varphi[a(U \circ T)f] \le V_\varphi[aA_M(Tf)] \le V_\varphi[aA_M A_L f].$$

The same result is clearly obtained when both U and T are *linear* operators; in this case $A_M := \int_{\mathbb{R}^+} H(t)\,dt$ and $A_L := \int_{\mathbb{R}^+} K(t)\,dt$. In particular when U is the linear operator of Riemann–Liouville we obtain an inequality concerning an estimate of Tf in terms of the fractional variation of a function $f \in L^1_{\text{loc}}(\mathbb{R}^+) \cap \text{Dom}\,T$.

Analogously, the number $V_\varphi[(U \circ T)f]$ may be interpreted as the (U, φ)-variation of Tf. In this respect, Theorem 6.2 gives an embedding result for T, with respect to the U-Musielak–Orlicz variation.

6.4 Dependence on a parameter: the space $\widetilde{BV}_\varphi$

In this section we extend the results given in Sections 6.2 and 6.3 to a more general situation when the φ-function generating the Musielak–Orlicz variation depends on a parameter. As before we will consider $I = \mathbb{R}^+$ as base measure space, provided with its Lebesgue measure. Let $\varphi : \mathbb{R}^+ \times \mathbb{R}^+_0 \to \mathbb{R}^+_0$ be a function such that $\varphi(t, \cdot)$ is a φ-function, for every $t \in \mathbb{R}^+$, and $\varphi(\cdot, u)$ is Lebesgue measurable for every $u \in \mathbb{R}^+_0$. For the sake of simplicity we denote by Φ the class of all these functions and when $\varphi(t, \cdot)$ is also convex, for every $t \in \mathbb{R}^+$, we write $\varphi \in \widetilde{\Phi}$.

We give now the generalized concept of variation. Let $D = \{t_i\}_{i=0,1,\dots,N} \subset \mathbb{R}^+$ a finite increasing sequence and let $s_i \in [t_{i-1}, t_i]$ for $i = 1, 2, \dots, N$. Then, for every $f \in \mathcal{X}$, we define the *generalized Musielak–Orlicz φ-variation of f* in $\mathbb{R}^+$ by means of the following formula:

$$\widetilde{V}_\varphi(f) \equiv \widetilde{V}_\varphi[f, I] = \sup_D \sum_{i=1}^N \varphi(s_i, |f(t_i) - f(t_{i-1})|)$$

where the supremum is taken over all the finite sequences $D = \{t_i\}_{i=0,1,\dots,N} \subset \mathbb{R}^+$ and the intermediate points $\{s_i\}_{i=1,\dots,N}$.

As for the Musielak–Orlicz variation, the functional $\widetilde{V_\varphi}$ is not a modular functional, but we can define a modular by putting

$$\widetilde{\rho}(f) = |f(a)| + \widetilde{V_\varphi}(f),$$

where a is a fixed point in I. Finally we denote by $\widetilde{BV_\varphi}(I)$ the corresponding modular space, called the space of functions with *bounded generalized Musielak–Orlicz φ-variation* in I.

Now, in order to obtain estimates in this setting, we need of the following growth condition on the function φ, called *s-boundedness condition* which represents a suitable modification of the boundedness condition for φ-functions depending on a parameter (see formula (1.9), Example 1.10), which takes into account of the variation functional.

A function $\varphi \in \Phi$ will be said *s-bounded* in $I = \mathbb{R}^+$, if there exist measurable functions $h : \mathbb{R}^+ \to \mathbb{R}_0^+$ and $l : \mathbb{R}^+ \times \mathbb{R}^+ \to \mathbb{R}$ and positive constants N_1, N_2 such that, for every finite, increasing sequence $\{s_0, s_1, \dots, s_N\} \subset \mathbb{R}^+$, there results

$$\varphi(t, u) \le N_1 h(z)\varphi(tz, N_2 u) + |\Delta(l(\cdot, z); [s_{i-1}, s_i])|$$

for every $t \in [s_{i-1}, s_i] \subset \mathbb{R}^+$, $i = 1, 2, \dots, N$, $z \in \mathbb{R}^+$, $u \in \mathbb{R}_0^+$ and where

$$\Delta(l(\cdot, z); [s_{i-1}, s_i]) = l(s_i, z) - l(s_{i-1}, z).$$

If $\varphi(t, u)$ is a convex function with respect to $u \in \mathbb{R}_0^+$, for every $t \in \mathbb{R}^+$, we will write $\varphi \in \widetilde{\Phi}$ and in the s-boundedness condition we may take $N_1 = 1$.

Example 6.2. Here we give a nontrivial class of functions $\varphi \in \Phi$, which satisfies the s-boundedness property. Take $\varphi(t, u) = t^\alpha \sigma(u)$, $\alpha \in \mathbb{R}$, $t \in \mathbb{R}^+$, where σ is a φ-function of u. Then if $\alpha = 0$, $\varphi(t, u) = \sigma(u)$, clearly satisfies the above condition for $N_1, N_2 \ge 1$, $h \ge 1$ and for arbitrary functions l or for suitable N_1, N_2, h and l. If $\alpha \ne 0$, it is sufficient to choose $l(t, z) = k\beta(z)$, for some constant $k \in \mathbb{R}_0^+$ and $\beta : \mathbb{R}^+ \to \mathbb{R}_0^+$, (i.e. l is a constant function with respect to $t \in \mathbb{R}^+$). In this case the s-boundedness condition is satisfied with $\Delta(l(\cdot, z); [a, b]) \equiv 0$, for any interval $[a, b] \subset \mathbb{R}^+$, $h(z) = z^{-\alpha}$, for each $t, z \in \mathbb{R}^+$, and $u \in \mathbb{R}_0^+$ and $N_1 = N_2 = 1$.

We will use the following notation: given a function $l : \mathbb{R}^+ \times \mathbb{R}^+ \to \mathbb{R}$, we will denote by $V_l(z) = V[l(\cdot, z)]$ the variation of $l(t, z)$ with respect to $t \in \mathbb{R}^+$, for every fixed $z \in \mathbb{R}^+$. Let $K \in \mathcal{K}$ and T be the corresponding integral operator of Section 6.2.

In order to establish an estimate for Tf we will use the following extension of the condition (6.4).

Let $\varphi : \mathbb{R}^+ \times \mathbb{R}_0^+ \to \mathbb{R}_0^+$, $\varphi \in \widetilde{\Phi}$, $\psi : \mathbb{R}_0^+ \to \mathbb{R}_0^+$, $\psi \in \Psi$; *there exist measurable functions* $\gamma : \mathbb{R}^+ \times \mathbb{R}_0^+ \to \mathbb{R}_0^+$, $\gamma \in \Phi$, *and* $v : \mathbb{R}^+ \to \mathbb{R}_0^+$ *such that*

$$\varphi(s, \psi(t, u)) \leq v(t)\gamma(s, u) \tag{6.6}$$

for every $t, s \in \mathbb{R}^+$, $u \in \mathbb{R}_0^+$.

We give some examples.

Example 6.3. A class of functions satisfying condition (6.6) is given by $\varphi(s, u) = b(s)u^\alpha$ with $\alpha \geq 1$, being $b : \mathbb{R}^+ \to \mathbb{R}_0^+$ a measurable function. Now, if we take $\psi(t, u) = g(t)r(u)$ with $g : \mathbb{R}^+ \to \mathbb{R}_0^+$ measurable and $r : \mathbb{R}_0^+ \to \mathbb{R}_0^+$, $r \in \Psi$, then there results $\varphi \in \widetilde{\Phi}$, $\psi \in \Psi$, and $\varphi(s, \psi(t, u)) = b(s)g^\alpha(t)r^\alpha(u) = v(t)\gamma(s, u)$, where $v(t) = g^\alpha(t)$ and $\gamma(s, u) = b(s)r^\alpha(u) \in \Phi$.

Now we are ready to establish that the operator T maps $\operatorname{Dom} T \cap \widetilde{BV}_\gamma(\mathbb{R}^+)$ into $\widetilde{BV}_\varphi(\mathbb{R}^+)$.

Theorem 6.3. *Let* $K \in \mathcal{K}$, $\varphi \in \widetilde{\Phi}$ *and let* $\gamma \in \Phi$ *be a function satisfying the s-boundedness condition, with constants* N_1, N_2 *and functions* $e : \mathbb{R}^+ \to \mathbb{R}_0^+$, $d : \mathbb{R}^+ \times \mathbb{R}^+ \to \mathbb{R}$. *Moreover we suppose that* γ *satisfies condition* (6.6) *and* $V_d(t) < +\infty$ *a.e.* $t \in \mathbb{R}^+$. *Let a be a constant with* $0 < a \leq \frac{1}{A_L}$ *and suppose that*

i) $0 < A_L := \int_{\mathbb{R}^+} L(t)\, dt < +\infty$,

ii) $C_L^e := \int_{\mathbb{R}^+} L(t)e(t)v(t)\, dt < +\infty$,

iii) $D_L^{V_d} := \int_{\mathbb{R}^+} L(t)v(t)V_d(t)\, dt < +\infty$.

Then, if $f \in \operatorname{Dom} T$, *there results*

$$\widetilde{V}_\varphi[a(Tf)] \leq N_1 \frac{C_L^e}{A_L} \widetilde{V}_\gamma[aA_L N_2 f] + \frac{D_L^{V_d}}{A_L}. \tag{6.7}$$

Proof. We may suppose $\widetilde{V}_\gamma[aA_L N_2 f] < +\infty$. Let $D = \{s_i\}_{i=0,1,\dots,N}$ be a finite increasing sequence in $\mathbb{R}^+$ and let $\xi_i \in [s_{i-1}, s_i]$ for $i = 1, 2, \dots, N$. For fixed but arbitrary $i \in \{1, 2, \dots, N\}$ we have that

$$|(Tf)(s_i) - (Tf)(s_{i-1})| \leq \int_{\mathbb{R}^+} L(t)\psi(t, |f(s_i t) - f(s_{i-1} t)|)\, dt.$$

Hence, by monotonicity of $\varphi(t,u)$ with respect to $u \in \mathbb{R}_0^+$, Jensen's inequality, concavity of $\psi(t,\cdot)$, condition (6.6) and s-boundedness of γ, we may write

$$\begin{aligned}
&\sum_{i=1}^{N} \varphi(\xi_i, a|(Tf)(s_i) - (Tf)(s_{i-1})|) \\
&\quad \le \sum_{i=1}^{N} \varphi(\xi_i, a \int_{\mathbb{R}^+} L(t)\psi(t, |f(s_i t) - f(s_{i-1}t)|)\, dt) \\
&\quad \le \frac{1}{A_L} \int_{\mathbb{R}^+} L(t) \sum_{i=1}^{N} \varphi(\xi_i, aA_L\psi(t, |f(s_i t) - f(s_{i-1}t)|))\, dt \qquad (+) \\
&\quad \le \frac{1}{A_L} \int_{\mathbb{R}^+} L(t)v(t) \sum_{i=1}^{N} \gamma(\xi_i, aA_L|f(s_i t) - f(s_{i-1}t)|)\, dt \\
&\quad \le \frac{1}{A_L} \int_{\mathbb{R}^+} L(t)v(t) \sum_{i=1}^{N} [N_1 e(t)\gamma(t\xi_i, aA_L N_2|f(s_i t) - f(s_{i-1}t)|)]\, dt \\
&\quad\quad + \frac{1}{A_L} \int_{\mathbb{R}^+} L(t)v(t) \sum_{i=1}^{N} |\Delta(d(\cdot, t); [s_{i-1}, s_i])|\, dt \\
&\quad \le N_1 \frac{1}{A_L} \int_{\mathbb{R}^+} L(t)e(t)v(t)\widetilde{V}_\gamma[aA_L N_2 f]\, dt + \frac{1}{A_L} \int_{\mathbb{R}^+} L(t)v(t)V_d(t)\, dt \\
&\quad = N_1 \frac{C_L^e}{A_L} \widetilde{V}_\gamma[aA_L N_2 f] + \frac{D_L^{V_d}}{A_L}.
\end{aligned}$$

Hence the assertion follows by the arbitrariness of D and $\{\xi_i\}$. □

Remark 6.3. a) Theorem 6.3 represents an extension of Theorem 6.1. Indeed if we put $\varphi(t,u) = \varphi(u)$ and $\gamma(t,u) = \gamma(u)$ i.e. φ and γ do not depend on the parameter $t \in \mathbb{R}^+$, then the s-boundedness condition for the function γ is satisfied in particular with $e(t) \equiv 1$, $N_1 = N_2 = 1$ and $\Delta(d(\cdot,t);[a,b]) \equiv 0$ for any interval $[a,b] \subset I$, and condition (6.6) becomes condition (6.4). Hence $C_L^e \equiv A_L^v$, $D_L^{V_d} \equiv 0$ and inequality (6.7) becomes the assertion of Theorem 6.1.

b) In case when the kernel K satisfies a strongly-Lipschitz condition, we obtain

$$\widetilde{V}_\varphi[a(Tf)] \le \frac{C_L^h}{A_L} \widetilde{V}_\varphi[aA_L N_2 f] + \frac{D_L^{V_l}}{A_L}$$

being $C_L^h := \int_{\mathbb{R}^+} L(t)h(t)\, dt$, $D_L^{V_l} := \int_{\mathbb{R}^+} L(t)V_l(t)\, dt$, with $0 < C_L^h, D_L^{V_l} < +\infty$, and $h : \mathbb{R}^+ \to \mathbb{R}_0^+$ and $l : \mathbb{R}^+ \times \mathbb{R}^+ \to \mathbb{R}$ are the functions of the s-boundedness of $\varphi \in \widetilde{\Phi}$. In this case we have that T maps Dom $T \cap \widetilde{BV}_\varphi(\mathbb{R}^+)$ in $\widetilde{BV}_\varphi(\mathbb{R}^+)$.

c) We remark here that the s-boundedness condition may be replaced by the following slighter one.

There exist measurable functions $h : \mathbb{R}^+ \to \mathbb{R}_0^+$ and $F : \mathbb{R}^+ \to \mathbb{R}_0^+$ and positive constants N_1, N_2 such that, for every finite increasing sequence $\{s_i\}_{i=0,1,\dots,N} \subset \mathbb{R}^+$, there results

$$\sum_{i=1}^{N} \varphi(\xi_i, u_i) \leq N_1 h(t) \sum_{i=1}^{N} \varphi(\xi_i t, N_2 u_i) + F(t) \qquad (++)$$

for every $N \in \mathbb{N}$, for every $\xi_i \in [s_{i-1}, s_i], i = 1, \dots, N$ and for every $u_i, i = 1, 2, \dots, N, u_i \in \mathbb{R}_0^+$.

This condition can be obtained by the previous one, on putting $t = \xi_i, u = u_i$, for every fixed $i = 1, 2, \dots, N$ and passing to the sum with i running from 1 to N; here $F(t) = V_l(t)$. With this condition, (6.7) of Theorem 6.3 holds with $D_L^{V_d} = D_L^G := \int_{\mathbb{R}^+} L(t)G(t)\,dt$, where $G(t) = V_d(t)$. Condition (++) is exactly what we need in order to obtain the above estimate. But the original s-boundedness condition is more readable and, as we remarked before, it is similar to the form of the boundedness condition used in the theory of Musielak–Orlicz spaces.

6.5 A superposition theorem in $\widetilde{BV}_\varphi$

In this section we will study inequalities for the composition of two nonlinear integral operators, with the same reasonings as in Section 6.3.

In order to do this, we have to modify assumption ($\star$) of Section 6.3, due to the fact that we deal now with functions φ, σ, λ depending on a parameter. So, we will assume the following condition.

Let $\varphi \in \widetilde{\Phi}$, $\psi, \theta \in \Psi$; there exist measurable functions $\sigma, \lambda : \mathbb{R}^+ \times \mathbb{R}_0^+ \to \mathbb{R}_0^+$ with $\sigma \in \widetilde{\Phi}$ and $\lambda \in \Phi$, and (measurable) $v_1, v_2 : \mathbb{R}^+ \to \mathbb{R}_0^+$, such that the following conditions are satisfied

$$\varphi(s, \theta(t, u)) \leq v_1(t)\sigma(s, u), \quad \sigma(s, \psi(t, u)) \leq v_2(t)\lambda(s, u) \qquad (\star\star)$$

for every $s, t \in \mathbb{R}^+, u \in \mathbb{R}_0^+$.

Examples of functions satisfying ($\star\star$) are similar to those given in Section 6.3 for the condition ($\star$), taking into account that here the functions φ, σ and λ depend on the parameter $t \in \mathbb{R}^+$.

Now we are ready to establish that, given two nonlinear integral operators U and T, as defined in Section 6.3, $(U \circ T)$ maps $\mathrm{Dom}T \cap \widetilde{BV}_\lambda(\mathbb{R}^+)$ in $\widetilde{BV}_\varphi(\mathbb{R}^+)$. Namely we can formulate the following

Theorem 6.4. *Let $H, K \in \mathcal{K}$, with H (M, θ)-Lipschitz and let $\varphi \in \widetilde{\Phi}$, $\psi, \theta \in \Psi$; let moreover $\sigma \in \widetilde{\Phi}$ and $\lambda \in \Phi$ be two s-bounded functions satisfying condition ($\star\star$). We denote by $m : \mathbb{R}^+ \to \mathbb{R}_0^+$, $p : \mathbb{R}^+ \times \mathbb{R}^+ \to \mathbb{R}$ and $Q \in \mathbb{R}^+$ respectively the functions and the constant of the s-boundedness of $\sigma \in \widetilde{\Phi}$, while by $n : \mathbb{R}^+ \to \mathbb{R}_0^+$,*

$q : \mathbb{R}^+ \times \mathbb{R}^+ \to \mathbb{R}$ and $P_1, P_2 \in \mathbb{R}^+$ respectively the functions and the constants of the s-boundedness of $\lambda \in \Phi$. Let moreover $V_p(t)$ and $V_q(t) < +\infty$, a.e. $t \in \mathbb{R}^+$ and let A_L with $0 < A_L < +\infty$ and A_M with $0 < A_M < +\infty$ be the constants as in Theorem 6.2. Moreover we suppose

i) $C_M^m := \int_{\mathbb{R}^+} M(t)m(t)v_2(t)\,dt < +\infty$,

ii) $D_M^{V_p} := \int_{\mathbb{R}^+} M(t)v_2(t)V_p(t)\,dt < +\infty$,

iii) $C_L^n := \int_{\mathbb{R}^+} L(t)n(t)v_1(t)\,dt < +\infty$,

iv) $D_L^{V_q} := \int_{\mathbb{R}^+} L(t)v_1(t)V_q(t)\,dt < +\infty$,

and let a be such that $0 < a < \min\left\{\frac{1}{A_M}, \frac{1}{A_L A_M Q}\right\}$. Then if $f \in \operatorname{Dom} T$, there results

$$\begin{aligned} \widetilde{V_\varphi}[a(U \circ T)f] &\le \frac{C_M^m}{A_M}\widetilde{V_\sigma}[aA_M Q(Tf)] + \frac{D_M^{V_p}}{A_M} \\ &\le \frac{C_M^m}{A_M}P_1\frac{C_L^n}{A_L}\widetilde{V_\lambda}[aP_2 Q A_M A_L f] + \frac{C_M^m}{A_M}\frac{D_L^{V_q}}{A_L} + \frac{D_M^{V_p}}{A_M}. \end{aligned} \tag{6.8}$$

Proof. The proof runs on similar lines as in Theorem 6.2; the details are left to the reader. □

Remark 6.4. a) Following similar reasoning as in Remark 6.2, we may obtain, as a particular case, an estimate with a strongly-Lipschitz condition and hence also an estimate for the composition of two linear integral operators.

b) Theorem 6.4 is an extension of Theorem 6.2 in Section 6.3: indeed if we take $\varphi(t,u) = \varphi(u), \sigma(t,u) = \sigma(u)$ and $\lambda(t,u) = \lambda(u)$, then the s-boundedness conditions for the functions σ and λ are satisfied with $m(t) = n(t) = 1$, $Q = P_1 = P_2 = 1$ and $\Delta(p(\cdot,t); [a,b]) = \Delta(q(\cdot,t); [a,b]) \equiv 0$ for any interval $[a,b] \subset I$ and condition $(\star\star)$ becomes condition $(\star)$.

6.6 The problem of convergence in φ-variation

In Chapters 3 and 4 we discussed modular approximation theorems to f, for families of nonlinear integral operators $\{T_w\}$ of convolution type, or with kernel satisfying some homogeneity assumptions, where f belongs to modular function spaces, in which the generating modular functional satisfies some suitable assumptions, like monotonicity, finiteness, absolute continuity, etc. These assumptions play an important role in the convergence properties of the modulus of continuity (see Sec. 2.2). As we remarked before, the modular ρ generating the space $BV_\varphi(I)$ or $\widetilde{BV}_\varphi(I)$, does not satisfy any of the above assumptions. This makes the problem of modular convergence of $T_w f$ towards f, in the space $BV_\varphi(I)$ or $\widetilde{BV}_\varphi(I)$ very difficult. We limit ourselves to consider the space $BV_\varphi(I)$.

A family of functions $f_w : I \to \mathbb{R}$, $f_w \in BV_\varphi(I)$, is said to be *convergent in φ-variation to $f \in BV_\varphi$, for $w \to w_0$, $w \in \mathbb{R}$, $w_0 \in \overline{\mathbb{R}}$,* if there is a constant $\lambda > 0$ such that

$$\lim_{w \to w_0} V_\varphi[\lambda(f_w - f)] = 0.$$

In the following, for the sake of simplicity, we will assume that w runs over the positive real axis $]0, +\infty[$ and $w_0 = +\infty$.

In this frame a suitable modulus of continuity is defined by

$$\omega(f, \delta) = \sup_{|h|<\delta} V_\varphi(\tau_h f - f),$$

where $(\tau_h f)(s) = f(s + h)$, is the translation operator, in the case when the interval $I = [a, b]$, and the involved functions are extended by periodicity $b-a$ outside $[a, b]$. When I is the group of positive real numbers, it is more convenient to use the dilation operator $(\mathcal{E}_h f)(s) = f(hs)$, and the modulus of continuity takes now the form

$$\omega(f, \delta) = \sup_{|h-1|<\delta} V_\varphi(\mathcal{E}_h f - f).$$

In both cases, it is well known that $\omega(f, \delta) \to 0$, as $\delta \to 0$, if and only if the function f satisfies a special continuity assumption, called φ-absolute continuity, (see [163] for the translation operator, [30], [177] for the dilation operator). So, unlike the theory of modular function spaces described in Section 2.2, the convergence property of the modulus of continuity ω doest not hold for *every* function $f \in BV_\varphi(I)$.

We quote here the definition of φ-absolute continuity. Let us suppose first that $I = [a, b]$. In this case we say that $f : I \to \mathbb{R}$ is *φ-absolutely continuous in I*, if there is $\lambda > 0$ such that the following condition holds: for every $\varepsilon > 0$ there is $\delta > 0$ such that

$$\sum_{i=1}^{N} \varphi[\lambda|f(\beta_i) - f(\alpha_i)|] < \varepsilon,$$

for all finite sets of non-overlapping intervals $[\alpha_i, \beta_i] \subset [a, b]$, $i = 1, 2, \ldots, N$ such that

$$\sum_{i=1}^{N} \varphi(\beta_i - \alpha_i) < \delta.$$

We denote the space of all φ-absolutely continuous functions with $AC_\varphi(I)$. If I is an open, or unbounded interval, for example, $I = \mathbb{R}^+$, we will say that f is *locally*

φ-absolutely continuous in I if $f \in AC_\varphi(J)$, for every compact interval $J \subset I$, with an absolute constant $\lambda > 0$ (see Section 6.7 below).

It is well known that $AC_\varphi(I)$ is a subspace of $BV_\varphi(I)$, when I is a bounded interval (see [163]).

The above considerations, suggest that a modular convergence theorem of type

$$V_\varphi[\lambda(T_w f - f)] \to 0, \quad w \to +\infty, \tag{6.9}$$

is possible only when f belongs to suitable subspaces of $BV_\varphi(I)$. In particular, if $f \in AC_\varphi(I)$, or locally, it is natural to think that the above convergence is true. This is really true for some class of *linear* integral operators. Let us mention two results in this direction.

For families of linear integral operators of convolution type, $\{T_w\}$, in which the kernel satisfies a classical singularity assumption, (6.9) holds for every function $f \in AC_\varphi(I)$, where $I = [a, b]$, and the involved functions are extended by periodicity $b - a$ outside I. Here the φ-function satisfies some further suitable assumption, (see e.g. [163], [188]).

A recent general result like (6.9), was given in [177], in which $I = \mathbb{R}^+$, and the dilation operator $\mathcal{E}_h$ is used. Here, $\{T_w\}$ is a family of linear integral operators, whose (singular) kernel satisfies a general homogeneity condition (as in Section 4.1, in the case of $R = 1$), with respect to a weight function $\zeta : I \to \mathbb{R}_0^+$. In fact, the operators here studied are the linear counterparts of those considered in Chapter 4.1. The modular approximation theorem established in [177], gives a convergence result similar to (6.9) for every $f \in \mathcal{X}$ such that an appropriate auxiliary function, linked to f, is locally absolutely continuous in I. Namely, putting $g(t) = t\zeta(t)f(t)$, there results

$$V_\varphi[\lambda(T_w f - g)] \to 0, \quad w \to +\infty$$

whenever g is locally absolutely continuous in I.

Another important result given in [177] concerns with the *rate of modular approximation* in suitable subspaces of $BV_\varphi(I)$ which represent the "variational" analogy of the modular Lipschitz classes (see Chapter 3).

We report here the definition of this class. Let Γ be the class of all measurable functions $\gamma : I \to \mathbb{R}_0^+$ such that $\gamma(1) = 0$, and $\gamma(s) \neq 0$ for $s \neq 1$. For a fixed $\gamma \in \Gamma$, we define the class:

$$\mathrm{Lip}_\gamma(V_\varphi) = \{f \in BV_\varphi(I) : \exists \nu > 0 : V_\varphi[\nu(\mathcal{E}_s f - f)] = \mathcal{O}(\gamma(s)), \text{ as } s \to 1\},$$

where, as usual, for any two functions $f, g \in X$, $f(s) = \mathcal{O}(g(s))$ as $s \to 1$ means that there are constants $C > 0, \delta > 0$ such that $|f(s)| \leq C|g(s)|$, for $s \in [1 - \delta, 1 + \delta]$.

In this respect, the result in [177] states that, under suitable assumptions, similar to those used in Chapter 4, if $f \in \mathcal{X}$ is such that $g(t) = t\zeta(t)f(t) \in \mathrm{Lip}_\gamma(V_\varphi)$, then for sufficiently small $\lambda > 0$ we have:

$$V_\varphi[\lambda(T_w f - g)] = \mathcal{O}(\xi(w^{-1})), \quad \text{as } w \to +\infty.$$

This is the "state of the art" for what concerns the convergence in the space $BV_\varphi(I)$.

For families of *nonlinear* integral operators a convergence theorem was recently proved in [25], in the particular case when the kernel functions K_w are of type $K_w(t,u) = L_w(t)H_w(u)$, for every $t \in \mathbb{R}^+$, and $u \in \mathbb{R}$, and by using a suitable notion of singularity for the family of kernel functions K_w. We will discuss these results in the next section. For more general nonlinear integral operators there are not known convergence theorems in $BV_\varphi(I)$.

Finally we remark that in the case of the space $\widetilde{BV}_\varphi(I)$, also the linear case is an open problem.

6.7 A convergence theorem in BV_φ for Mellin-type nonlinear integral operators

In this section, without restrictions, we will consider the closed interval $\mathbb{R}_0^+ = [0, +\infty[$, and we will denote again by $\mathcal{X}$ the space of all Lebesgue measurable functions $f : \mathbb{R}_0^+ \to \mathbb{R}$.

Moreover we will consider here the class Φ' of φ-functions, $\varphi : \mathbb{R}_0^+ \to \mathbb{R}_0^+$, such that

i) φ is a convex function on $\mathbb{R}_0^+$,

ii) $u^{-1}\varphi(u) \to 0$ as $u \to 0^+$.

We begin with some further definitions.

We will say that a family of functions $\{f_w\}_{w>0} \subset \mathcal{X}$ is of *equibounded φ-variation*, if it is of bounded φ-variation, uniformly with respect to $w > 0$.

Now we recall the following result about φ-variation, which we will use in the following (for a proof see [163]):

j) if $f_1, f_2, \dots, f_n \in \mathcal{X}$, then

$$V_\varphi\Big[\sum_{i=1}^{n} f_i\Big] \le \frac{1}{n}\sum_{i=1}^{n} V_\varphi[nf_i].$$

Let $\varphi, \eta \in \Phi'$ be fixed. We will say that a function $f : \mathbb{R}_0^+ \to \mathbb{R}$ is *locally (φ, η)-absolutely continuous* if there is a $\lambda > 0$ such that the following property holds: for every $\varepsilon > 0$ and for every bounded interval $J \subset \mathbb{R}_0^+$ there is $\delta > 0$ such that for any finite collection of non-overlapping intervals $[a_i, b_i] \subset J$, $i = 1, 2, \dots, N$, with $\sum_{i=1}^{N} \varphi(b_i - a_i) < \delta$ there results

$$\sum_{i=1}^{N} \eta(\lambda|f(b_i) - f(a_i)| < \varepsilon. \tag{6.10}$$

If $\eta = \varphi$ in the above property, we will say that f is locally φ-absolutely continuous (see [163], [153], [177]), and we will denote by $AC^{\varphi}_{\text{loc}}(\mathbb{R}^+_0)$ the class of all these functions.

We will say that a family of functions $\{f_w\}_{w>0}$ is *locally equi (φ, η)-absolutely continuous* if there is $\lambda > 0$ such that for every $\varepsilon > 0$ and every bounded interval $J \subset \mathbb{R}^+_0$, we can choose a $\delta > 0$ for which the local (φ, η)-absolute continuity of f_w holds uniformly with respect $w > 0$. For $\eta = \varphi$ we will say that $\{f_w\}_{w>0}$ is locally equi φ-absolutely continuous.

Let now $\mathcal{K}$ be the class of all the functions $K : \mathbb{R}^+_0 \times \mathbb{R} \to \mathbb{R}$ of the form

$$K(t, u) = L(t)H(u), \quad t \in \mathbb{R}^+_0, \ u \in \mathbb{R},$$

where $L \in L^1(\mathbb{R}^+_0)$, $L \geq 0$ and $H : \mathbb{R} \to \mathbb{R}$ is a function satisfying a Lipschitz condition of type

$$|H(u) - H(v)| \leq \psi(|u - v|), \quad u, v \in \mathbb{R}, \tag{6.11}$$

where $\psi : \mathbb{R}^+_0 \to \mathbb{R}^+_0$ is a function with the following properties:

(1) $\psi(0) = 0$, $\psi(u) > 0$ for $u > 0$,

(2) ψ is continuous and nondecreasing.

We will denote again by Ψ the class of all functions ψ satisfying the above conditions.

Let $\mathbb{K} = \{K_w\}_{w>0}$ be a set of functions from $\mathcal{K}$, $K_w(t, u) = L_w(t)H_w(u)$, $w > 0, t \in \mathbb{R}^+_0, u \in \mathbb{R}$. We will say that $\mathbb{K}$ is *singular* in $BV_\varphi(\mathbb{R}^+_0)$, if the following assumptions hold:

(K.1) there exists $A > 0$, such that $0 < \|L_w\|_1 = A_w \leq A$ for every $w > 0$,

(K.2) for every $\delta \in]0, 1[$, we have

$$\lim_{w \to +\infty} \int_{|1-t|>\delta} L_w(t)\, dt = 0,$$

(K.3) putting $G_w(u) = H_w(u) - u$, for every $u \in \mathbb{R}$, $w > 0$, there exists $\lambda > 0$ such that

$$V_\varphi[\lambda G_w, J] \to 0, \quad \text{as } w \to +\infty,$$

for every bounded interval $J \subset I$.

Example 6.4. For every $n \in \mathbb{N}$, let

$$K_n(t, u) = L_n(t)H_n(u), \quad t \in \mathbb{R}^+_0, \ u \in \mathbb{R},$$

where

$$H_n(u) = \begin{cases} n \log(1 + u/n), & 0 \leq u < 1 \\ nu \log(1 + 1/n), & u \geq 1, \end{cases}$$

where we extend in odd-way the definition of H_n for $u < 0$; moreover $\{L_n\}_{n\in\mathbb{N}}$ is a classical kernel with the mass concentrated at 1, i.e.

$$\int_0^{+\infty} L_n(t)\,dt = 1, \qquad \text{for every } n \in \mathbb{N},$$

with the property (K.2). It is easy to show that

$$|H_n(u) - H_n(v)| \leq |u - v|, \qquad \text{for every } u, v \in \mathbb{R}, \text{ and } n \in \mathbb{N}$$

and, for every $u \geq 0$, we have

$$|G_n(u)| = |H_n(u) - u| = \begin{cases} u - n\log(1 + u/n), & 0 \leq u < 1 \\ u[1 - n\log(1 + 1/n)], & u \geq 1. \end{cases}$$

Then $|G_n(u)|$ is increasing on $\mathbb{R}_0^+$. If $\varphi : \mathbb{R}_0^+ \to \mathbb{R}_0^+$ is a convex function, using Proposition 1.03 in [163], we have, for every interval $J = [0, M]$,

$$V_\varphi[G_n, J] = \varphi(|G_n(M) - G_n(0)|) \to 0, \qquad \text{as } n \to +\infty.$$

Analogously, by the definition of H_n for $u < 0$, we have $V_\varphi[G_n, [-M, 0]] \to 0$, as $n \to +\infty$.

Before we formulate the following lemmas, we recall the definition of convergence in φ-variation (see Section 6.6).

We recall that a sequence $(f_w)_{w\in\mathbb{R}^+} \in BV_\varphi$ is convergent in φ-variation to $f \in BV_\varphi$ if there exists a $\lambda > 0$ such that $V_\varphi[\lambda(f_w - f)] \to 0$ as $w \to +\infty$.

Moreover we will use the following relation between the functions φ, ψ and η, where φ and η are two φ-functions, with η not necessarily convex, and $\psi \in \Psi$.

We say that the triple $\{\varphi, \eta, \psi\}$ is *properly directed*, if the following condition holds (for similar assumptions see [143]; compare with the definition of Section 1.4): for every $\lambda > 0$, there exists a constant C_λ such that

$$\varphi(C_\lambda \psi(u)) \leq \eta(\lambda u), \qquad \text{for every } u \geq 0. \tag{6.12}$$

Now we start formulating the following lemma.

Lemma 6.1. *Let $f : \mathbb{R}_0^+ \to \mathbb{R}$ be a locally (φ, η)-absolutely continuous function. Let $\{H_w\}_{w>0}$ be a class of functions satisfying (6.11) for a fixed $\psi \in \Psi$ and for every $w > 0$ and let us assume that the triple $\{\varphi, \eta, \psi\}$ is properly directed. Then the family $\{H_w \circ f\}_{w>0}$ is locally equi φ-absolutely continuous.*

Proof. Let $\lambda > 0$ be a constant for which the definition of (φ, η)-absolute continuity of f holds and let $0 < \mu \leq C_\lambda$, being C_λ the constant in (6.12). Since f is locally

(φ, η)-absolutely continuous, for a fixed interval $J \subset \mathbb{R}_0^+$ and $\varepsilon > 0$ there is a $\delta > 0$ such that (6.10) holds for any finite collection of intervals $I_i = [a_i, b_i], i = 1, 2, \dots N$, with $\sum_{i=1}^N \varphi(b_i - a_i) < \delta$. For such a family $\{I_i\}$, we have

$$\begin{aligned}\sum_{i=1}^N &\varphi(\mu|(H_w \circ f)(b_i) - (H_w \circ f)(a_i)|)\\ &\le \sum_{i=1}^N \varphi(C_\lambda \psi(|f(b_i) - f(a_i)|))\\ &\le \sum_{i=1}^N \eta(\lambda|f(b_i) - f(a_i)|) < \varepsilon.\end{aligned}$$

□

Lemma 6.2. *Let f be a locally φ-absolutely continuous function such that $f \in BV_\varphi(\mathbb{R}_0^+)$. Let $\{H_w\}_{w>0}$ be a family of functions $H_w : \mathbb{R} \to \mathbb{R}$ such that* (K.3) *holds. Then there is $\lambda > 0$ such that the following property holds: for every $\varepsilon > 0$ and every bounded interval $[0, b] \subset \mathbb{R}_0^+$, there exist $\overline{w} > 0$ and a step function $\nu : \mathbb{R}_0^+ \to \mathbb{R}$ such that*

$$V_\varphi(\lambda(H_w \circ f - \nu), [0, b]) < \varepsilon$$

uniformly with respect to $w \ge \overline{w}$.

Proof. Let $[0, b] \subset \mathbb{R}_0^+$ be a fixed bounded interval. From Lemma 1 in [177] (see also Theorem 2.21 of [163]), there is a $\lambda > 0$ such that, for a fixed $\varepsilon > 0$ there is a partition $D = \{\tau_0 = 0, \tau_1, \dots, \tau_n = b\}$ of the interval $[0, b]$, such that the step function $\nu : \mathbb{R}_0^+ \to \mathbb{R}$, defined by

$$\nu(t) = \begin{cases} f(\tau_{i-1}), & \tau_{i-1} \le t < \tau_i,\ i = 1, \dots m \\ f(b), & t \ge b \end{cases}$$

satisfies

$$V_\varphi(2\lambda(f - \nu), [0, b]) < \varepsilon/2.$$

Let $D = \{t_0, t_1, \dots, t_n\}$ be an arbitrary partition of $[0, b]$ with $t_0 < t_1 < \dots < t_n$. We have

$$\begin{aligned}\sum_{i=1}^n &\varphi(\lambda|H_w(f(t_i)) - \nu(t_i) - \{H_w(f(t_{i-1})) - \nu(t_{i-1})\}|)\\ &\le \frac{1}{2}\sum_{i=1}^n \varphi(2\lambda|H_w(f(t_i)) - f(t_i) - \{H_w(f(t_{i-1})) - f(t_{i-1})\}|)\\ &+ \frac{1}{2}\sum_{i=1}^n \varphi(2\lambda|f(t_i) - \nu(t_i) - \{f(t_{i-1}) - \nu(t_{i-1})\}|) = I_1 + I_2.\end{aligned}$$

Since $f \in BV_\varphi(\mathbb{R}_0^+)$, f is bounded, i.e. there is a constant $M > 0$ such that $|f(t)| \leq M$. Putting $J = [-M, M]$, we have

$$I_1 \leq \frac{1}{2} V_\varphi(2\lambda G_w, J).$$

Thus using (K.3) we can take $\lambda > 0$ such that $I_1 \leq \varepsilon/2$ for sufficiently large $w > 0$. The assertion follows being $I_2 \leq \frac{1}{2} V_\varphi(2\lambda(f - \nu), [0, b]) < \varepsilon/2$. □

Lemma 6.3. *Let $f \in BV_\eta(\mathbb{R}_0^+)$ and $\{H_w\}$ be a family of functions $H_w : \mathbb{R} \to \mathbb{R}$ satisfying* (6.11). *Let us suppose that the triple $\{\varphi, \eta, \psi\}$ is properly directed. Then the family $\{H_w \circ f\}$ is of equibounded φ-variation on every interval $I^* \subset \mathbb{R}_0^+$.*

Proof. Let $D = \{t_0, t_1, \dots t_n\} \subset I^*$ be fixed and let $\lambda > 0$. For $0 < \mu \leq C_\lambda$, C_λ being the constant in (6.12), we have

$$\sum_{i=1}^{n} \varphi(\mu|(H_w \circ f)(t_i) - (H_w \circ f)(t_{i-1})|) \leq \sum_{i=1}^{n} \varphi(C_\lambda \psi(|f(t_i) - f(t_{i-1}|)).$$

Now, by (6.12) we have

$$\sum_{i=1}^{n} \varphi(\mu|(H_w \circ f)(t_i) - (H_w \circ f)(t_{i-1})|) \leq \sum_{i=1}^{n} \eta(\lambda|f(t_i) - f(t_{i-1}|) \leq V_\eta(\lambda f, I^*),$$

and so the assertion follows. □

For any $z \in \mathbb{R}^+$, we will put

$$\tau_z f(s) = f(sz),$$

for every $f : \mathbb{R}_0^+ \to \mathbb{R}$ and $s \in \mathbb{R}_0^+$. Using the above lemmas, we show the following theorem

Theorem 6.5. *Let φ, η be fixed, and let $f : \mathbb{R}_0^+ \to \mathbb{R}$ be a locally φ-absolutely continuous function, such that $f \in BV_{\varphi+\eta}(\mathbb{R}_0^+)$. Let $\{H_w\}$ be a family of functions $H_w : \mathbb{R} \to \mathbb{R}$ satisfying* (K.3) *and* (6.11) *for a fixed $\psi \in \Psi$. Let us assume that the triple $\{\varphi, \eta, \psi\}$ is properly directed. Then for every $\lambda > 0$ there exist a constant $\mu > 0$ and $\overline{w} > 0$ such that*

$$\lim_{z \to 1} V_\varphi[\mu(\tau_z(H_w \circ f) - (H_w \circ f))] = 0$$

uniformly with respect to $w \geq \overline{w}$.

Proof. Let $g_w = H_w \circ f$, for $w > 0$. Since $f \in BV_\eta(\mathbb{R}_0^+)$, from Lemma 1 of [177], given $\varepsilon > 0$ there is $c > 0$ and $\lambda_0 > 0$ such that $V_\eta(\lambda f, [c, +\infty[) < \varepsilon$, for every $0 < \lambda \leq \lambda_0$. From Lemma 6.3, there exists a constant $\mu > 0$ so small that

$$V_\varphi(4\mu g_w, [c, +\infty[) \leq V_\eta(\lambda f, [c, +\infty[) < \varepsilon$$

uniformly with respect to $w > 0$. Let us choose constants d, b with $d > b > c$ and let ν be the step function on $[0, d]$ given in Lemma 6.2. Let now z be such that $c/b < z < \min\{d/b, b/c\}$. By convexity of φ, and property j), for every z sufficiently near to 1, we have now, for sufficiently small $\mu > 0$,

$$\begin{aligned}
&V_\varphi[\mu(\tau_z g_w - g_w)] \\
&\quad \leq \frac{1}{2}\{V_\varphi[2\mu(\tau_z g_w - g_w), [0, b]] + V_\varphi[2\mu(\tau_z g_w - g_w), [b, +\infty[]\} \\
&\quad \leq \frac{1}{2} V_\varphi[2\mu(\tau_z g_w - g_w), [0, b]) \\
&\qquad + \frac{1}{4}\{V_\varphi(4\mu\tau_z g_w, [b, +\infty[) + V_\varphi(4\mu g_w, [b, +\infty[)\} \\
&\quad \leq \frac{1}{2} V_\varphi[2\mu(\tau_z g_w - g_w), [0, b]] + \frac{1}{2} V_\eta(\lambda f[c, +\infty[) \\
&\quad \leq \frac{1}{2} V_\varphi[2\mu(\tau_z g_w - g_w), [0, b]] + \varepsilon.
\end{aligned}$$

The first inequality comes from a classical property of φ-variation (see [163], Proposition 1.17).

Now we consider the interval $I^* = [0, b]$. We have, for sufficiently small $\mu > 0$,

$$\begin{aligned}
&V_\varphi[2\mu(\tau_z g_w - g_w), I^*] \\
&\quad \leq \frac{1}{3}\{V_\varphi[6\mu\tau_z(g_w - \nu), I^*] + V_\varphi[6\mu(\nu - g_w), I^*] + V_\varphi[6\mu(\tau_z \nu - \nu), I^*]\} \\
&\quad \leq \frac{1}{3}\{2V_\varphi[6\mu(g_w - \nu), [0, d]] + V_\varphi[6\mu(\tau_z \nu - \nu), [0, d]]\} \\
&\quad = I_1 + I_2.
\end{aligned}$$

From Lemma 6.2, $I_1 \leq \varepsilon/2$, while as in Theorem 1 in [30], we have $I_2 \leq \varepsilon/2$. Thus the assertion follows. □

Let $\mathbb{K} = \{K_w(t, u)\}_{w>0}$ be a singular kernel in $BV_\varphi(\mathbb{R}_0^+)$, where, as before, $K(t, u) = L_w(t) H_w(u)$ for $t \in \mathbb{R}_0^+$, $u \in \mathbb{R}$.

We will study approximation properties of the family of nonlinear integral operators $\mathbb{T} = \{T_w\}$ defined by

$$(T_w f)(s) = \int_0^{+\infty} K_w(t, f(st))\, dt = \int_0^{+\infty} L_w(t) H_w(f(st))\, dt, \quad s \in \mathbb{R}_0^+,$$

where $f \in \operatorname{Dom} \mathbb{T}$ is the class of all measurable functions $f : \mathbb{R}_0^+ \to \mathbb{R}$ such that $T_w f$ is well defined as a Lebesgue integral for every $s \in \mathbb{R}_0^+$. Let us remark here that if the function f is such that $(H_w \circ f) \in L^1(\mathbb{R}_0^+)$, or if $f \in L^\infty(\mathbb{R}_0^+)$, then $f \in \operatorname{Dom} \mathbb{T}$. So in particular, if f is of bounded φ-variation, where φ is an arbitrary φ-function, $f \in \operatorname{Dom} \mathbb{T}$.

Let now $\varphi \in \Phi'$ and η be two φ-functions, with φ convex and η not necessarily convex, such that the triple $\{\varphi, \eta, \psi\}$ is properly directed. Then, by Theorem 6.1, if $f \in BV_\eta(\mathbb{R}_0^+)$, then $T_w f$ is of bounded φ-variation for every $w > 0$ (see also [143]).

We have the following

Theorem 6.6. *Let $f \in AC_{\text{loc}}^\varphi(\mathbb{R}_0^+) \cap BV_{\varphi+\eta}(\mathbb{R}_0^+)$ and let us assume that the triple $\{\varphi, \eta, \psi\}$ is properly directed. Let $\mathbb{K} = \{K_w\} \subset \mathcal{K}$ be a singular family of kernel functions in $BV_\varphi(\mathbb{R}_0^+)$. Then there exists a constant $\mu > 0$ such that*

$$\lim_{w \to +\infty} V_\varphi[\mu(T_w f - f)] = 0.$$

Proof. First of all we remark that $T_w f - f \in BV_\varphi(\mathbb{R}_0^+)$. We can assume that $A_w = 1$, for every $w > 0$, where A_w are the constants given in (K.1). Let $\lambda > 0$ be such that $V_\eta(\lambda f) < +\infty$, and let $\mu > 0$ so small that $4\mu \le C_\lambda$ and

$$\lim_{z \to 1} V_\varphi[2\mu(\tau_z(H_w \circ f) - (H_w \circ f))] = 0,$$

uniformly with respect to sufficiently large $w > 0$ (Theorem 6.5).

Let $D = \{s_0, s_1, \dots, s_N\} \subset \mathbb{R}_0^+$ be a finite increasing sequence and let μ be sufficiently small. We have

$$\begin{aligned}
&\sum_{i=1}^N \varphi[\mu|(T_w f)(s_i) - (T_w f)(s_{i-1}) - f(s_i) + f(s_{i-1})|] \\
&\quad = \sum_{i=1}^N \varphi\Big[\mu\Big|\int_0^{+\infty} L_w(t)[H_w(f(s_i t)) - H_w(f(s_i)) \\
&\qquad\qquad + H_w(f(s_i)) - f(s_i) - H_w(f(s_{i-1}t)) \\
&\qquad\qquad + H_w(f(s_{i-1})) - H_w(f(s_{i-1}) + f(s_{i-1})]\, dt\Big|\Big] \\
&\quad \le \frac{1}{2}\sum_{i=1}^N \int_0^{+\infty} L_w(t)\varphi[2\mu|(H_w(f(s_i t)) - H_w(f(s_i))) \\
&\qquad\qquad - (H_w(f(s_{i-1}t)) - H_w(f(s_{i-1})))|]\, dt \\
&\qquad\qquad + \frac{1}{2}\sum_{i=1}^N \int_0^{+\infty} L_w(t)\varphi[2\mu|(H_w(f(s_i)) - f(s_i)) \\
&\qquad\qquad - (H_w(f(s_{i-1})) - f(s_{i-1}))|]\, dt \\
&\quad = I_1 + I_2.
\end{aligned}$$

Now given $\delta \in]0,1[$, we write

$$\begin{aligned} I_1 &\leq \frac{1}{2}\sum_{i=1}^{N}\left\{\int_{|1-t|<\delta} + \int_{|1-t|>\delta}\right\} \\ &\qquad L_w(t)\varphi[2\mu|(H_w(f(s_i t)) - H_w(f(s_i))) \\ &\qquad\qquad - (H_w(f(s_{i-1}t)) - H_w(f(s_{i-1})))|]\,dt \\ &= I_1^1 + I_1^2. \end{aligned}$$

Next,

$$I_1^1 \leq \frac{1}{2}\int_{1-\delta}^{1+\delta} L_w(t)V_\varphi(2\mu[\tau_t(H_w \circ f) - (H_w \circ f)])\,dt$$

and so, for sufficiently small $\delta \in]0,1[$ we have $I_1^1 \leq \varepsilon$, uniformly with respect to $w > 0$.

Now, by property j),

$$I_1^2 \leq \frac{1}{4}\int_{|1-t|>\delta} L_w(t)V_\varphi(4\mu(H_w \circ f))\,dt \leq \frac{1}{4}V_\eta(\lambda f)\int_{|1-t|>\delta} L_w(t)\,dt,$$

and so, from (K.2), $I_1^2 \to 0$, as $w \to +\infty$.

Finally, we estimate I_2. We have

$$I_2 \leq \frac{1}{2}\int_0^{+\infty} L_w(t)V_\varphi(2\mu G_w)\,dt = \frac{1}{2}V_\varphi(2\mu G_w).$$

But since f is bounded, there is a constant $M > 0$, such that $|f(t)| \leq M$ for every $t \in \mathbb{R}_0^+$. Putting $J = [-M, M]$, we apply the singularity assumption (K.3) and we obtain $I_2 \to 0$ as $w \to +\infty$.

The proof is now complete. □

6.8 Bibliographical notes

The classical concept of (Jordan) variation of a function $f : [a,b] \to \mathbb{R}$ was firstly generalized in 1924 by N. Wiener [208], who introduced the notion of quadratic variation of f. This work was followed by a series of papers of L. C. Young [209], [210] and E. R. Love and L. C. Young [138], and by E. R. Love [135], where generalized variation with power of order $p \geq 1$, the respective generalized absolute continuity and application to Riemann–Stieltjes integral and to Fourier series are given. The definition of φ-variation given here, was introduced by L. C. Young [210] and further developed by J. Musielak and W. Orlicz [163], in the case when φ is not dependent on the parameter. In the same paper also the concept of absolute continuity with respect to φ is introduced and its important connections with the convergence of the modulus of continuity are studied. Other contribution to the theory of φ-variation were given by H. Herda [116], R. Lésniewicz and W. Orlicz [134] and J. Musielak [149], in which spaces of sequences of finite φ-variation are considered.

The results given in Sections 6.2, 6.3 were proved by I. Mantellini and G. Vinti in [143]. The linear case was considered in [31] for general linear operators with homogeneous kernel of degree α, and for superposition of two linear integral operators, in which the homogeneity assumptions are given in a generalized sense(ζ-homogeneity). In [31] some applications to fractional calculus are also given, in particular embedding theorems are obtained for linear operators with homogeneous kernel, with respect to the fractional φ-variation, introduced there by using the Riemann–Liouville integral of f. In the classical case ($\varphi(u) = u$), the fractional variation was studied by several authors, in connection with "shape-preserving" properties of linear operators (see e.g., [212], [99]). Multidimensional versions of these results are available for linear integral operators in [31]. Here a concept of Vitali φ-variation is introduced in a natural way.

In a series of papers (see [105]–[110]), S. Gnilka investigated the properties of the generalized variation, in which the function φ depends on a parameter. Also a generalized concept of absolute continuity was introduced and studied.

G. Vinti [201] introduced a multidimensional version of the generalized φ-variation, when φ depends on a parameter, following the approach of the Vitali variation and here the notion of s-boundedness for φ-functions depending on a parameter is also introduced. In this setting, some estimate of linear operators with homogeneous kernel are obtained, and some application to fractional calculus is studied.

Results about convergence of linear integral operators with respect to φ-variation were given firstly in [163] for sequences of linear integral operators of convolution type and then this result was extended by J. Szelmeczka [188] for filtered families of such operators. In [30], by using the dilation operator $\mathcal{E}_h$, this result was extended to moment type operators, acting on functions defined on $\mathbb{R}^+$. There, the "multiplicative" version of the modulus of continuity is given, using the dilation operator $\mathcal{E}_h$ and its connections with φ-absolute continuity are studied.

The moment kernel satisfies a homogeneity condition with degree -1 (if we consider Lebesgue measure). Based on this remark, S. Sciamannini and G. Vinti [177] gave some convergence theorems in φ-variation, for a general class of linear operators with ζ-homogeneous kernels. Moreover, by introducing the corresponding Lipschitz classes, the rate of convergence is also studied. Recently, in [178] a convergence result for more general nonlinear integral operators of Volterra type has been considered.

The convergence results of Section 6.7 are given in [25].

A related problem is a "weak" form of the convergence in φ-variation for a family of integral operators, namely a relation of type

$$V_\varphi[T_w f] \to V_\varphi[f], \tag{6.13}$$

for $f \in BV_\varphi(I)$ or $f \in \widetilde{BV}_\varphi(I)$.

This problem takes its origin from classical problems of Calculus of Variations, through the work of L. Tonelli [192], T. Radó [173], L. Cesari [82], [83], E. Baiada [5], E. Baiada and G. Cardamone [6], E. Baiada and C. Vinti [7], C. Vinti [196], [197], [198], [199].

In particular in [197],[198] the Tonelli–Cesari variation is considered in the multidimensional case, and results like (6.13) are obtained, for various classes of linear integral operators (also non-convolution). Applications to convergence in perimeter, area and length are also given there. The true definition of perimeter, introduced by E. De Giorgi in 1954 (see [86], [87]), is essentially given by a convergence in (distributional) variation, similar to (6.13). Distributional versions of the results given in [173], [197], [199], were obtained by C. Goffman and J. Serrin [111], M. Boni [48], [49], E. Michener [146], C. Bardaro and D. Candeloro [11], [12], C. Bardaro [10], C. Bardaro and G. Vinti [28], [33]. In these papers a concept of $\mathcal{F}$-variation is used, in the multidimensional case, where $\mathcal{F}$ is a sublinear functional. Applications to the Serrin integral of Calculus of Variations are also given. For similar results, in the setting of Weierstrass Integrals, see also [80], [81].

The general problem of "weak" convergence in φ-variation is still open.

Chapter 7
Application to nonlinear integral equations

7.1 An embedding theorem

Let (Ω, Σ, μ) be a measure space with a σ-finite, complete measure, endowed with a commutative operation $+$ from $\Omega \times \Omega$ to Ω.

We are going to investigate nonlinear integral equations generated by the convolution-type operator (3.1), i.e. equations of the form

$$\int_{\Omega} K(t, f(t+s))\, d\mu(t) = f(s) + g(s) \tag{7.1}$$

for μ-almost all $s \in \Omega$, where $g \in L^0(\Omega)$ is given and $f \in L^0(\Omega)$ is the unknown function. In order that the left-hand side of this equation makes sense, we suppose that the space $L^0(\Omega)$ is invariant and K is a Carathéodory kernel function. Under these assumptions, we have $K(\cdot, f(\cdot+s)) \in L^0(\Omega)$ for every $s \in \Omega$ (see Section 3.1). In order to be able to apply the Fubini–Tonelli theorem, we should know that $K(t, f(t+s))$ is a measurable function on $\Omega \times \Omega$. Let Σ_0 be the smallest σ-algebra of subsets $C \subset \Omega \times \Omega$ such that $A \times B \in \Sigma_0$ whenever $A, B \in \Sigma$, and let Σ_π be any σ-algebra of subsets of $\Omega \times \Omega$ such that $\Sigma_0 \subset \Sigma_\pi$. We denote by μ_0 the product measure on Σ_0, i.e. $\mu_0(A \times B) = \mu(A)\mu(B)$ for $A, B \in \Sigma$, and we denote by μ_π any extension of the measure μ_0 from Σ_0 to Σ_π. We denote by $\mathcal{K}_\pi$ the class of all Carathédory functions $K : \Omega \times \mathbb{R} \to \mathbb{R}$ such that the function $\widetilde{K} : \Omega \times \Omega \to \mathbb{R}$ defined by $\widetilde{K}(s, t) = K(t, f(t+s))$ is Σ_π-measurable for every $f \in L^0(\Omega)$. Functions $K \in \mathcal{K}_\pi$ will be called *Σ_π-regular Carathéodory functions.*

Example 7.1. Let μ be the Lebesgue measure on the σ-algebra Σ of all Lebesgue measurable subsets of $\Omega = \mathbb{R}$ and let μ^2 be the Lebesgue measure on the σ-algebra Σ^2 of all Lebesgue measurable subsets of $\mathbb{R}^2$. Let $+$ be a commutative operation on $\mathbb{R}$ such that the space $L^0(\mathbb{R})$ is invariant. We shall write $\sigma(s, t) = s+t$, and we suppose that σ is (Σ^2, Σ)-measurable, i.e., if $A \in \Sigma$ then $\sigma^{-1}(A) \in \Sigma^2$. Moreover, we suppose that μ^2 is *σ-absolutely continuous (σ-a.c.)* with respect to μ, i.e., if $A \in \Sigma$ and $\mu(A) = 0$, then $\mu^2(\sigma^{-1}(A)) = 0$. Let $K : \Omega \times \mathbb{R} \to \mathbb{R}$ be a Carathéodory kernel function. We shall show that K is Σ^2-regular, i.e., the function $\widetilde{K}(s, t) = K(t, f(t+s))$ is Σ^2-measurable for every $f \in L^0(\Omega)$. First, we shall prove it in the case when f has a bounded support, i.e., $f(t) = 0$ outside a compact interval $[a, b] \subset \mathbb{R}$. By Fréchet's theorem, there exists a sequence (f_n) of continuous functions on $[a, b]$ such that $f_n(t) \to f(t)$ μ-a.e. in $[a, b]$. Denoting by A the set of $t \in \Omega$ for which the sequence

$(f_n(t))$ does not converge to $f(t)$, we have $\mu(A) = 0$. Hence $\mu^2(\sigma^{-1}(A)) = 0$, by the σ-absolute continuity of μ^2 with respect to μ. If $(s,t) \in \sigma^{-1}(A)$, i.e., $\sigma(s,t) \in A$, then $f_n(\sigma(s,t)) \to f(\sigma(s,t))$ as $n \to +\infty$, whence $f_n(\sigma(s,t)) \to f(\sigma(s,t))$ as $n \to +\infty$ μ^2-a.e. in $\mathbb{R}^2$. If we know that $K(t, f_n((\sigma(s,t)))$ are Σ^2-measurable then also $K(t, f((\sigma(s,t)))$ is Σ^2-measurable, since the measure μ^2 is complete. So, we may limit ourselves to the case of continuous functions f. Let $f : \mathbb{R} \to \mathbb{R}$ be continuous on $\mathbb{R}$. Let (σ_i) be a sequence of simple integrable functions with respect to $(\mathbb{R}^2, \Sigma^2, \mu^2)$, convergent to the function $\sigma(s,t) = s+t$ for all $(s,t) \in \mathbb{R}^2$. This means that there exist pairwise disjoint sets $A_1^{(i)}, A_2^{(i)}, \dots, A_{k_i}^{(i)} \in \Sigma^2$ of finite measure μ^2 and constants $c_1^{(i)}, c_2^{(i)}, \dots, c_{k_i}^{(i)} \in \mathbb{R}$ such that $\sigma_i(s,t) = \sum_{j=1}^{k_i} c_j^{(i)} \chi_{A_j^{(i)}}(s,t)$ converges to $\sigma(s,t)$ for all $s,t \in \mathbb{R}$. By continuity of f and $K(t,\cdot)$, and from the condition $K(t,0) = 0$, we obtain

$$K(t, f(\sigma(s,t))) = \lim_{i\to+\infty} K(t, f(\sigma_i(s+t))) = \lim_{i\to+\infty} \sum_{j=1}^{k_i} K(t, f(c_j^{(i)}))\chi_{A_j^{(i)}}(s,t).$$

But the function at the right-hand side of the above equality is Σ^2-measurable. Thus $K(t, f(\sigma(s,t)))$ is Σ^2-measurable. Now we omit the assumption that f has bounded support. Let $f \in L^0(\mathbb{R})$ be arbitrary. We put $f_n(t) = f(t)$ for $t \in [-n,n]$ and $f_n(t) = 0$ otherwise. Then $f_n(t) \to f(t)$ as $n \to +\infty$ for $t \in \mathbb{R}$. By continuity of $K(t,\cdot)$, we have $K(t, f_n(\sigma(s,t))) \to K(t, f(\sigma(s,t)))$ as $n \to +\infty$ for all $s,t \in \mathbb{R}$. As we have already proved, $K(t, f_n(\sigma(s,t)))$ are Σ^2-measurable for $n = 1,2,\dots,$ since $f_n \in L^0(\mathbb{R})$. Thus, $K(t, f(\sigma(s,t)))$ is Σ^2-measurable for every $f \in L^0(\mathbb{R})$. Consequently, K is a Σ^2-regular Carathéodory function. □

Let us still remark that the usual addition $+$ in $\mathbb{R}$ is Σ^2-measurable and μ^2 is σ-a.c., since we have then $\sigma^{-1}(A) = \{(s,t) \in \mathbb{R}^2 : s+t \in A\}$ for any $A \subset \mathbb{R}$.

It is obvious that if K is a Σ_π-regular Carathéodory function, then its absolute value $|K|$ defined by $|K|(t,u) = |K(t,u)|$ is also a Σ_π-regular Carathéodory kernel. Thus, the integral

$$\int_{\Omega\times\Omega} |K(t, f(t+s))|\, d\mu_\pi(s,t)$$

exists for every $f \in L^0(\Omega)$. By the Fubini–Tonelli theorem, the integral

$$\int_\Omega |K(t, f(t+s))|\, d\mu(t)$$

exists for μ-almost all $s \in \Omega$ and is a Σ-measurable function of the variable s in Ω. Obviously, this integral may be infinite. We are going to formulate a theorem showing that for functions f from some modular spaces $L_\rho^0(\Omega)$, we have $\int_\Omega |K(t, f(t+s))|\, d\mu(t) < +\infty$ μ-a.e. on Ω, and that $L_\rho^0(\Omega) \subset \operatorname{Dom} T$ (see Section 1.1).

Theorem 7.1. *Let the following assumptions be satisfied.*

(a) *ρ is a J-quasiconvex modular on $L^0(\Omega)$, subbounded with respect to the operation $+$,*

(b) *K is a Σ_π-regular, $(L,\psi)_0$-Lipschitz Carathéodory kernel function, where $L:\Omega\to\mathbb{R}_0^+$ is a Σ-measurable function such that $0\neq L\in L^1(\Omega)$, and $\psi:\Omega\times\mathbb{R}_0^+\to\mathbb{R}_0^+$ satisfies the conditions: $\psi(\cdot,u)$ is Σ-measurable for all $u\geq 0$, $\psi(t,:)$ is continuous, concave and nondecreasing for every $t\in\Omega$, $\psi(t,0)=0$, $\psi(t,u)>0$ for $u>0$, $\psi(t,u)\to+\infty$ as $u\to+\infty$ for all $t\in\Omega$,*

(c) *$L(\cdot)\psi(\cdot,1)\in L^1(\Omega)$, and*

$$\int_\Omega L(t)\psi(t,1)|f(t)|\,d\mu(t)<+\infty$$

for every $f\in L^0_\rho(\Omega)$.

Then $L^0_\rho(\Omega)\subset \mathrm{Dom}\,T$.

Proof. First we show that if $f\in L^0_\rho(\Omega)$ then $\int_\Omega |K(t,f(t+s))|\,d\mu(t)<+\infty$ for $s\in\Omega$. Since ψ is concave, we have for $u\geq 1$

$$u\psi(t,1)=u\psi\left(t,\frac{1}{u}u\right)\geq u\frac{1}{u}\psi(t,u)$$

for every $t\in\Omega$. For a fixed $s\in\Omega$ we put $A_s=\{t\in\Omega:|f(t+s)|\geq 1\}$, $B_s=\Omega\setminus A_s$. By the $(L,\psi)_0$-Lipschitz condition, we obtain

$$\begin{aligned}|K(t,f(t+s))| &\leq L(t)\psi(t,|f(t+s)|)\\ &= L(t)\psi(t,|f(t+s)|)\chi_{B_s}(t)+L(t)\psi(t,|f(t+s)|)\chi_{A_s}(t)\\ &\leq L(t)\psi(t,1)\chi_{B_s}(t)+L(t)\psi(t,1)|f(t+s)|\chi_{A_s}(t).\end{aligned}$$

Since K is Σ_π-regular, the function $|K(t,f(t+s))|$ is Σ_π-measurable in $\Omega\times\Omega$. Hence it follows

$$\begin{aligned}&\int_\Omega |K(t,f(t+s))|\,d\mu(t)\\ &\quad\leq \int_{B_s} L(t)\psi(t,1)\,d\mu(t)+\int_{A_s} L(t)\psi(t,1)|f(t+s)|\,d\mu(t)\\ &\quad\leq \int_\Omega L(t)\psi(t,1)\,d\mu(t)+\int_\Omega L(t)\psi(t,1)|f(t+s)|\,d\mu(t).\end{aligned}$$

Since $L(\cdot)\psi(\cdot,1)\in L^1(\Omega)$, the first of the integrals on the right-hand side of the above inequality is finite. In order to prove that the integral $\int_\Omega |K(t,f(t+s))|\,d\mu(t)<+\infty$

it is sufficient to show that the second integral on the right-hand side of this inequality is finite, too. However, since ρ is subbounded with respect to $+$, we have the inequality $\rho(\lambda f(\cdot + s)) \leq \rho(C\lambda f) + h(t)$ with some $C \geq 1$, $0 \leq h \in L^\infty(\Omega)$ and every $f \in L^0_\rho(\Omega)$, $s \in \Omega$, $\lambda > 0$. Hence if $0 < \lambda \leq 1$, we have

$$\rho(\lambda^2 f(\cdot + s)) \leq \lambda M \rho(\lambda M f(\cdot + s)) \leq \lambda M \rho(C\lambda M f) + \lambda M \|h\|_\infty,$$

where $M \geq 1$ is the constant in the definition of J-quasiconvexity of ρ. Since the right-hand side of this inequality tends to 0 as $\lambda \to 0^+$ for every $f \in L^0_\rho(\Omega)$, so $\rho(\lambda^2 f(\cdot + s)) \to 0$ as $\lambda \to 0^+$. Hence $f(\cdot + s) \in L^0_\rho(\Omega)$ for every $s \in \Omega$. By assumption (c), we have

$$\int_\Omega L(t)\psi(t, 1)|f(t + s)|\, d\mu(t) < +\infty$$

for every $s \in \Omega$. Thus, we proved that

$$\int_\Omega |K(t, f(t + s))|\, d\mu(t) < +\infty$$

μ-a.e. on Ω. In order to prove Σ-measurability of $(Tf)(s) = \int_\Omega K(t, f(t+s))\, d\mu(t)$, let us remark that if K is a Σ_π-regular, $(L, \psi)_0$-Lipschitz Carathéodory kernel function, then both the positive part K_+ and negative part K_- of K are also $(L, \psi)_0$-Lipschitz, Σ_π-regular Carathéodory kernel functions. Denoting

$$(T_+ f)(s) = \int_\Omega K_+(t, f(t + s))\, d\mu(t), \quad (T_- f)(s) = \int_\Omega K_-(t, f(t + s))\, d\mu(t)$$

we observe, applying the first part of the proof to T_+ and T_- in place of T, that the integrals

$$\int_\Omega K_+(t, f(t + s))\, d\mu(t), \quad \int_\Omega K_-(t, f(t + s))\, d\mu(t) \tag{7.2}$$

are finite for almost all $s \in \Omega$. Applying the Tonelli theorem to the functions $K_+(t, f(t + s))$ and $K_-(t, f(t + s))$ on the product $\Omega \times \Omega$, we see that both integrals (7.2) are Σ-measurable functions of $s \in \Omega$. Thus,

$$(Tf)(s) = \int_\Omega K_+(t, f(t + s))\, d\mu(t) - \int_\Omega K_-(t, f(t + s))\, d\mu(t)$$

is a Σ-measurable function on Ω. Consequently, $L^0_\rho(\Omega) \subset \operatorname{Dom} T$. □

Example 7.2. Let us suppose that $\varphi : \mathbb{R}^+_0 \to \mathbb{R}^+_0$ is an *N-function*, i.e. it is convex, $\varphi(0) = 0$, $\varphi(u) > 0$ for $u > 0$ and

$$\lim_{u \to 0^+} \frac{\varphi(u)}{u} = 0 \quad \text{and} \quad \lim_{u \to +\infty} \frac{\varphi(u)}{u} = +\infty.$$

The function $\varphi^* : \mathbb{R}_0^+ \to \mathbb{R}_0^+$, defined by the formula

$$\varphi^*(u) = \sup_{v>0}(uv - \varphi(v)),$$

called *conjugate to φ in the sense of Young*, is also an N-function (see e.g. [153]). Let us denote

$$\rho(f) = \int_\Omega \varphi(|f(t)|)\,d\mu(t), \quad \rho^*(f) = \int_\Omega \varphi^*(|f(t)|)\,d\mu(t)$$

for $f \in L^0(\Omega)$, (see Example 1.5 (b)). The modular spaces $L^\varphi(\Omega) = L^0_\rho(\Omega)$ and $L^{\varphi^*}(\Omega) = L^0_{\rho^*}$ are Orlicz spaces generated by the above modulars. From the definition of the function φ^* we obtain the Young inequality

$$uv \le \varphi(u) + \varphi^*(v),$$

for $u, v \ge 0$, immediately. In particular, we have for every $\lambda > 0$,

$$\lambda^2 L(t)\psi(t,1)|f(t)| \le \varphi(\lambda|f(t)|) + \varphi^*(\lambda L(t)\psi(t,1)).$$

Hence

$$\begin{aligned}\int_\Omega L(t)\psi(t,1)|f(t)|\,d\mu(t) &\le \frac{1}{\lambda^2}\int_\Omega \varphi(\lambda|f(t)|)\,d\mu(t)\\ &\quad + \frac{1}{\lambda^2}\int_\Omega \varphi^*(\lambda L(t)\psi(t,1))\,d\mu(t)\\ &= \frac{1}{\lambda^2}\rho(\lambda f) + \frac{1}{\lambda^2}\rho^*(\lambda L(\cdot)\psi(\cdot,1)).\end{aligned}$$

Since $f \in L^0_\rho(\Omega)$, we have $\rho(\lambda f) < +\infty$ for sufficiently small $\lambda > 0$. Now, let us suppose that $L(\cdot)\psi(\cdot,1) \in L^0_{\rho^*}(\Omega)$, then also $\rho^*(\lambda L(\cdot)\psi(\cdot,1)) < +\infty$ for sufficiently small $\lambda > 0$. This shows that $\int_\Omega L(t)\psi(t,1)|f(t)|\,d\mu(t) < +\infty$ for every $f \in L^0_\rho(\Omega)$, i.e. the assumption (c) in Theorem 7.1 is satisfied, if only $L(\cdot)\psi(\cdot,1) \in L^0_{\rho^*}(\Omega) = L^{\varphi^*}(\Omega)$. In this manner we proved the following corollary to Theorem 7.1.

Corollary 7.1. *Let φ and φ^* be a pair of N-functions, conjugate in the sense of Young. Let K be a Σ_π-regular, $(L,\psi)_0$-Lipschitz Carathéodory kernel function with functions L and ψ satisfying the assumption* (b) *of Theorem* 7.1. *Moreover, let $L(\cdot)\psi(\cdot,1) \in L^{\varphi^*}(\Omega)$ and let*

$$(Tf)(s) = \int_\Omega K(t, f(t+s))\,d\mu(t).$$

Then $L^\varphi(\Omega) \subset$ Dom T.

Let us remark that under the conditions of Theorem 7.1 or of Corollary 7.1, we have obviously $L^0_\rho(\Omega) \cap \text{Dom}\,T = L^0_\rho(\Omega)$. This means that in this case the condition $f \in L^0_\eta(\Omega) \cap \text{Dom}\,T$, resp. $f \in L^0_{\rho+\eta}(\Omega) \cap \text{Dom}\,\mathbb{T}$, in Theorems 3.2 and 3.3 may be replaced by $f \in L^0_\eta(\Omega)$ resp. $f \in L^0_{\rho+\eta}(\Omega)$.

7.2 Existence and uniqueness results via Banach's fixed point principle

In the next theorem we show that, under suitable assumptions, the convolution-type operator T is a contraction.

Theorem 7.2. *Let ρ be a monotone, J-quasiconvex modular on an invariant space $L^0_\rho(\Omega)$, strongly subbounded with respect to the operation $+$ in Ω. Let K be a Carathéodory kernel function satisfying the Lipschitz condition*

$$|K(t,u) - K(t,v)| \le L(t)|u - v| \quad \text{for } u, v \in \mathbb{R},\ t \in \Omega,$$

where $L : \Omega \to \mathbb{R}^+_0$, $0 \ne L \in L^1(\Omega)$. Then

(a) $T : L^0_\rho(\Omega) \cap \operatorname{Dom} T \to L^0_\rho(\Omega)$.

(b) $\rho(\lambda(Tf - Tg)) \le M\rho[\lambda CM\|L\|_1(f - g)]$ *for $f, g \in L^0_\rho(\Omega) \cap \operatorname{Dom} T$, and any $\lambda > 0$, where M and C are the constants from the definitions of J-quasiconvexity and strong subboundedness of ρ, respectively.*

Moreover, if we suppose ρ to be convex and J-convex, then

(c) $\|Tf - Tg\|_\rho \le C\|L\|_1\|f - g\|_\rho$ *for $f, g \in L^0_\rho(\Omega) \cap \operatorname{Dom} T$.*

Proof. (a) Applying the monotonicity of ρ, the Lipschitz condition with $v = 0$, J-quasiconvexity of ρ with a constant $M \ge 1$ and strong subboundedness of ρ with a constant $C \ge 1$, successively, we obtain for arbitrary $f \in L^0_\rho(\Omega) \cap \operatorname{Dom} T$ and $\lambda > 0$

$$\begin{aligned}
\rho(\lambda Tf) &\le \rho\left(\lambda \int_\Omega L(t)|f(t+\cdot)|\,d\mu(t)\right)\\
&= \rho\left(\int_\Omega p(t)\lambda\|L\|_1|f(t+\cdot)|\,d\mu(t)\right)\\
&\le M\int_\Omega p(t)\rho[\lambda M\|L\|_1|f(t+\cdot)|]\,d\mu(t)\\
&\le M\int_\Omega p(t)\rho(\lambda CM\|L\|_1 f)\,d\mu(t) = M\rho(\lambda CM\|L\|_1 f),
\end{aligned}$$

where $p(t) = L(t)/\|L\|_1$ for $t \in \Omega$. Since $f \in L^0_\rho(\Omega)$, we have $\rho(\lambda CM\|L\|_1 f) \to 0$ as $\lambda \to 0^+$. By the above inequality, there holds $\rho(\lambda Tf) \to 0$ as $\lambda \to 0^+$. Consequently, $Tf \in L^0_\rho(\Omega)$.

(b) We apply the monotonicity of ρ, the Lipschitz condition, J-quasiconvexity and strong subboundedness of ρ, successively, obtaining for $f, g \in L^0_\rho(\Omega) \cap \operatorname{Dom} T$

and $\lambda > 0$

$$\begin{aligned}\rho(\lambda(Tf - Tg)) &\le \rho\left(\lambda \int_\Omega L(t)|f(t+\cdot) - g(t+\cdot)|\,d\mu(t)\right)\\ &\le M \int_\Omega p(t)\rho[\lambda M\|L\|_1|f(t+\cdot) - g(t+\cdot)|]\,d\mu(t)\\ &\le M \int_\Omega p(t)\rho[\lambda C M\|L\|_1(f-g)]\,d\mu(t)\\ &= M\rho[\lambda C M\|L\|_1(f-g)].\end{aligned}$$

(c) From the inequality in (b) with $M = 1$ we obtain, taking $\lambda = 1/u$ for arbitrary $u > 0$ and supposing $f, g \in L^0_\rho(\Omega) \cap \operatorname{Dom} T$, the inequality

$$\rho\left(\frac{Tf - Tg}{u}\right) \le \rho\left(\frac{C\|L\|_1(f-g)}{u}\right).$$

Hence

$$\{u > 0 : \rho\left(\frac{C\|L\|_1(f-g)}{u}\right) \le 1\} \subset \{u > 0 : \rho\left(\frac{Tf - Tg}{u}\right) \le 1\}.$$

Consequently

$$\begin{aligned}\|Tf - Tg\|_\rho &= \inf\{u > 0 : \rho\left(\frac{Tf - Tg}{u}\right) \le 1\}\\ &\le \inf\{u > 0 : \rho\left(\frac{C\|L\|_1(f-g)}{u}\right) \le 1\}\\ &= C\|L\|_1\|f - g\|_\rho.\end{aligned}$$

□

Now, applying the Banach's fixed point principle, we are able to prove the following theorem.

Theorem 7.3. *Let the following assumptions be satisfied:*

(a) *ρ is a monotone, convex, J-convex modular on an invariant space $L^0(\Omega)$, strongly subbounded with respect to the operation $+$ in Ω with a constant $C \ge 1$,*

(b) *The modular space $L^0_\rho(\Omega)$ is complete with respect to the norm $\|\cdot\|_\rho$,*

(c) *$L : \Omega \to \mathbb{R}^+_0$ satisfies the condition $0 \ne L \in L^1(\Omega)$ and $\int_\Omega L(t)|f(t)|\,d\mu(t) < +\infty$ for every $f \in L^0_\rho(\Omega)$,*

(d) *K is a Σ_π-regular Carathéodory kernel function satisfying the Lipschitz condition $|K(t,u) - K(t,v)| \le L(t)|u - v|$ for $u, v \in \mathbb{R}$ and $t \in \Omega$, where $C\|L\|_1 < 1$.*

Then, for every $g \in L^0_\rho(\Omega)$, the integral equation

$$\int_\Omega K(t, f(t+s))\, d\mu(t) = f(s) + g(s), \quad \mu\text{-a.e. on } \Omega, \tag{7.3}$$

has a unique solution $f \in L^0_\rho(\Omega)$.

Proof. Let us observe that from the assumptions (a)–(d) it follows that there are satisfied the assumptions of both Theorems 7.1 and 7.2, with $\psi(t,u) = |u|$ for all $t \in \Omega$ and $u \ge 0$. By Theorem 7.1, there holds $L^0_\rho(\Omega) \subset \operatorname{Dom} T$. Hence, by Theorem 7.2 (a), $T : L^0_\rho(\Omega) \to L^0_\rho(\Omega)$. If we put $T_1 f = Tf - g$, for $g \in L^0_\rho(\Omega)$ then also $T_1 : L^0_\rho(\Omega) \to L^0_\rho(\Omega)$. Moreover, we have $T_1 f_1 - T_1 f_2 = Tf_1 - Tf_2$ for $f_1, f_2 \in L^0_\rho(\Omega)$. Hence, by Theorem 7.2 (c), we have

$$\|T_1 f_1 - T_1 f_2\|_\rho = \|Tf_1 - Tf_2\|_\rho \le C\|L\|_1 \|f_1 - f_2\|_\rho$$

for $f_1, f_2 \in L^0_\rho(\Omega)$, and since $C\|L\|_1 < 1$, T_1 is a contraction in $L^0_\rho(\Omega)$. Since $L^0_\rho(\Omega)$ is complete, we may apply Banach's fixed point principle. Thus, there exists a unique $f \in L^0_\rho(\Omega)$ such that $T_1 f = f$, i.e., $Tf = f + g$, which means that the equation (7.3) admits f as a unique solution. □

Example 7.3. Let the modulars ρ and ρ^* be defined as in Example 7.2, where φ and φ^* are N-functions, conjugate in the sense of Young. Then ρ satisfies all assumptions of Theorem 7.3 with constant $C = 1$. Moreover, the Orlicz space $L^0_\rho(\Omega) = L^\varphi(\Omega)$ is complete. Let us suppose that K is a Σ_π-regular Carathéodory kernel function satisfying the Lipschitz condition $|K(t,u) - K(t,v)| \le L(t)|u - v|$ for $u, v \in \mathbb{R}$, $t \in \Omega$ with a function $L : \Omega \to \mathbb{R}^+_0$ such that $0 \ne L \in L^1(\Omega) \cap L^{\varphi^*}(\Omega)$ and $\|L\|_1 < 1$. Then, by Example 7.2, we have $\int_\Omega L(t)|f(t)|\, d\mu(t) < +\infty$ for every $f \in L^\varphi(\Omega)$, and so there are satisfied all the assumptions of Theorem 7.3. Thus the equation (7.1) has a unique solution $f \in L^\varphi(\Omega)$.

Let us still remark that the condition $\|L\|_1 < 1$ may be omitted, if we consider in place of (7.1) the integral equation

$$\lambda \int_\Omega K(t, f(t+s))\, d\mu(t) = f(s) + g(s),$$

where $0 < \lambda < 1/\|L\|_1$, replacing the kernel function K by means of the kernel function λK. Finally, let us remark that if $0 < \mu(\Omega) < +\infty$, then $L^{\varphi^*}(\Omega) \subset L^1(\Omega)$, and so the assumption $0 \ne L \in L^{\varphi^*}(\Omega) \cap L^1(\Omega)$ may be replaced by $0 \ne L \in L^{\varphi^*}(\Omega)$. Indeed, if $0 < \mu(\Omega) < +\infty$, we obtain for every $\alpha > 0$

$$\varphi^*\left(\frac{1}{\mu(\Omega)} \int_\Omega \alpha |f(t)|\, d\mu(t)\right) \le \frac{1}{\mu(\Omega)} \int_\Omega \varphi^*(\alpha|f(t)|)\, d\mu(t),$$

by convexity of φ^*. If $f \in L^{\varphi^*}(\Omega)$, then the right-hand side of the above inequality is finite for sufficiently small $\alpha > 0$. Hence $\int_\Omega |f(t)|\,d\mu(t) < +\infty$, i.e. $f \in L^1(\Omega)$.

7.3 Existence results via Schauder's fixed point principle

We shall now investigate the possibility of an application of the following Schauder fixed point principle to the integral equation (7.1).

Theorem 7.4 (Schauder fixed point principle). *Let X be a Banach space with norm $\|\cdot\|$ and let V_0 be a nonempty, compact, convex subset of X. Let $T : V_0 \to V_0$ be a continuous map of V_0 into itself. Then there exists a point $x_0 \in V_0$ such that $Tx_0 = x_0$.*

For the proof, see e.g. [95].

Applying this theorem, we prove the following

Corollary 7.2. *Let X be a Banach space with norm $\|\cdot\|$ and let V be a nonempty, closed, convex subset of X. Let $T : V \to V$ be a continuous map of V into itself such that the image TV of V is conditionally compact in X. Then there exists a point $x_0 \in V$ such that $Tx_0 = x_0$.*

Proof. Since the closure $\overline{TV}$ of the set TV is compact, so, by Mazur's theorem (see [95]) its closed convex hull $\overline{\mathrm{conv}(\overline{TV})}$ is also compact. Since $T : V \to V$, we have $TV \subset V$ and so $\overline{TV} \subset \overline{V} = V$, because V is closed and hence $\overline{\mathrm{conv}(\overline{TV})} \subset \overline{\mathrm{conv} V} = V$. Let us write $V_0 = \overline{\mathrm{conv}(\overline{TV})}$, then V_0 is a nonempty, compact, convex subset of X. The inclusion above shows that $V_0 \subset V$. Moreover, $TV \subset V_0$. Hence $TV_0 \subset V_0$, and so T maps V_0 into itself. Since T is continuous in V, it is also continuous in V_0.

Thus applying Theorem 7.4, we obtain that there exists an $x_0 \in V_0$ such that $Tx_0 = x_0$. □

We shall apply this corollary in the case when $X = L^0_\rho(\Omega)$ or $X = E^0_\rho(\Omega)$, where ρ is a convex modular in $L^0(\Omega)$ and $E_\rho(\Omega)$ is the set of *finite elements* of $L^0_\rho(\Omega)$, i.e., of functions $f \in L^0(\Omega)$ such that $\rho(\lambda f) < +\infty$ for all $\lambda > 0$. We have $E^0_\rho(\Omega) \subset L^0_\rho(\Omega)$ and we equip both these spaces with the norm $\|\cdot\|_\rho$, generated by the modular ρ. It is easily seen that if the space $L^0_\rho(\Omega)$ is complete with respect to the norm $\|\cdot\|_\rho$, then $E^0_\rho(\Omega)$ is complete with respect to the same norm, too. Indeed, let (f_n) be a Cauchy sequence in $E^0_\rho(\Omega)$. Then it is also a Cauchy sequence in $L^0_\rho(\Omega)$, and so there is an $f \in L^0_\rho(\Omega)$ such that $\|f_n - f\|_\rho \to 0$ as $n \to +\infty$. We have to show that $f \in E^0_\rho(\Omega)$. By Theorem 1.2, we have $\rho(\lambda(f_n - f)) \to 0$ as $n \to +\infty$ for every $\lambda > 0$. There holds

$$\rho(\lambda f) \leq \frac{1}{2}\rho(2\lambda(f - f_n)) + \frac{1}{2}\rho(2\lambda f_n)$$

for $n = 1, 2, \dots$ and for every $\lambda > 0$. From the above inequality it follows that for any $\lambda > 0$ there exists an index n such that $\rho(2\lambda(f_n - f)) < +\infty$. Since $f_n \in E^0_\rho(\Omega)$, so $\rho(2\lambda f_n) < +\infty$ for every $\lambda > 0$. Hence $\rho(\lambda f) < +\infty$ for every $\lambda > 0$. Thus, $f \in E^0_\rho(\Omega)$.

We shall prove the following

Theorem 7.5. *Let ρ be a monotone, convex, J-convex modular on $L^0_\rho(\Omega)$, strongly subbounded with a constant $C \geq 1$ with respect to the operation $+$ in Ω. Let K be an $(L, \psi)_0$-Lipschitz Carathéodory kernel function satisfying the assumptions* (b) *and* (c) *from Theorem* 7.1. *Moreover, let us suppose the triple $\{\rho, \psi, \rho\}$ to be properly directed with C_λ satisfying the condition $\lambda^{-1}C_\lambda \geq a > 0$ for any $\lambda > 0$. We denote by X any of two spaces $L^0_\rho(\Omega)$, $E^0_\rho(\Omega)$ and we put $V_\rho = \{f \in X : \|f\|_\rho \leq 1\}$. Let $(Tf)(s) = \int_\Omega K(t, f(t+s))\,d\mu(t)$ and $T_1 f = Tf - g$, where $g \in X$ and $\|g\|_\rho = \theta < 1$. Finally, let $\|L\|_1 \leq aC^{-1}(1-\theta)$. Then T_1 maps V_ρ into itself.*

Proof. First, let us remark that by Theorem 7.1, $X \subset L^0_\rho(\Omega) \subset \operatorname{Dom} T$. Applying the monotonicity of ρ, the $(L, \psi)_0$-Lipschitz condition and J-convexity of ρ, we obtain for arbitrary $\alpha > 0$

$$\begin{aligned}\rho(\alpha Tf) &\leq \rho\left(\alpha \int_\Omega |K(t, f(t+\cdot))|\,d\mu(t)\right)\\ &\leq \rho\left(\int_\Omega p(t)\alpha\|L\|_1\psi(t, |f(t+\cdot)|)\,d\mu(t)\right)\\ &\leq \int_\Omega p(t)\rho[\alpha\|L\|_1\psi(t, |f(t+\cdot)|)]\,d\mu(t).\end{aligned}$$

By the assumption that $\{\rho, \psi, \rho\}$ is properly directed for all $\lambda > 0$, taking $\alpha\|L\|_1 \leq C_\lambda$, we obtain

$$\rho[\alpha\|L\|_1\psi(t, |f(t+\cdot)|)] \leq \rho(\lambda|f(t+\cdot)|).$$

Hence, by the strong subboundedness of ρ with a constant $C \geq 1$, we obtain

$$\begin{aligned}\rho(\alpha Tf) &\leq \int_\Omega p(t)\rho(\lambda|f(t+\cdot)|)\,d\mu(t)\\ &\leq \int_\Omega p(t)\rho(\lambda Cf)\,d\mu(t) = \rho(\lambda Cf).\end{aligned}$$

Taking $\alpha = C_\lambda/\|L\|_1$, we thus obtain

$$\rho\left(\frac{C_\lambda}{\|L\|_1}Tf\right) \leq \rho(\lambda Cf)$$

for $\lambda > 0$. But $\lambda^{-1}C_\lambda \geq a$ for $\lambda > 0$, so from the last inequality we obtain

$$\rho\left(\frac{a}{\|L\|_1}\lambda Tf\right) \leq \rho(\lambda Cf)$$

for $\lambda > 0$. If we put $\lambda = 1/u$, we get

$$\rho\left(\frac{aTf}{u\|L\|_1}\right) \leq \rho(u^{-1}Cf)$$

for $u > 0$. Hence we obtain, similarly as in the proof of Theorem 7.2 (c), the inequality

$$\|Tf\|_\rho \leq \frac{C}{a}\|L\|_1\|f\|_\rho. \tag{7.4}$$

Since $\|L\|_1 \leq aC^{-1}(1-\theta)$, we get

$$\|Tf\|_\rho \leq (1-\theta)\|f\|_\rho,$$

if $f \in X$. Now, supposing $f \in V_\rho$, we have $\|f\|_\rho \leq 1$ and so

$$\|T_1 f\|_\rho \leq (1-\theta)+\theta = 1.$$

This shows that $T_1 : V_\rho \to V_\rho$. □

Let us remark that if the $(L, \psi)_0$-Lipschitz condition holds with the function $\psi(u) = u$, then the triple $\{\rho, \psi, \rho\}$ is properly directed with $C_\lambda = \lambda$, and the condition $\lambda^{-1}C_\lambda \geq a > 0$ is satisfied with $a = 1$.

From Corollary 7.2 and Theorem 7.5 we immediately obtain the following statement.

Theorem 7.6. *Let the assumptions of Theorem 7.5 be satisfied, and let the space $L^0_\rho(\Omega)$ be complete with respect to the norm $\|\cdot\|_\rho$. Let us suppose further that the operator*

$$Tf(s) = \int_\Omega K(t, f(t+s))\,d\mu(t)$$

is continuous on the ball V_ρ and the image TV_ρ is conditionally compact in X, then the integral equation (7.1) *has a solution $f \in V_\rho$.*

Remark 7.1. Let us assume that $(\Omega, +)$ is a locally compact and σ-compact group, endowed with its Haar measure μ on the σ-algebra of its Borel subsets. Since the modular ρ is monotone, the normed space $L^0_\rho(\Omega)$ is a preideal space . In this case there are known conditions for continuity of T and conditional compactness of $TV_\rho(X)$, also in the more general case of an Urysohn operator T (see [193], Theorem 2.1).

In case of a general modular ρ on $L^0_\rho(\Omega)$ we do not know any necessary and sufficient conditions in order that a set $A \subset X$ be conditionally compact in X, where $X = L^0_\rho(\Omega)$ or $X = E^0_\rho(\Omega)$. We shall quote the results in the case of a modular ρ defined by (1.4) from Example 1.5 (c). Proofs may be found e.g. in [153], Theorem 9.12.

Let $\Omega \subset \mathbb{R}^n$ be a Lebesgue measurable set and let $d\mu = dt$ be the Lebesgue measure in the σ-algebra Σ of all Lebesgue measurable subsets of Ω. Let Φ be the class of all functions $\varphi : \Omega \times \mathbb{R}_0^+ \to \mathbb{R}_0^+$ which are measurable with respect to the first variable for every value of the second variable, and are convex φ-functions of the second variable with respect to every $t \in \Omega$. The modular ρ will be defined by formula (1.4), i.e., $\rho(f) = \int_\Omega \varphi(t, |f(t)|)\, dt$ for every Lebesgue measurable function f on Ω (see Example 1.5 (c)). We say that $\varphi \in \Phi$ is *locally integrable* in Ω, if for every set $A \in \Sigma$ of finite measure and every $u \geq 0$ there holds $\int_A \varphi(t, u)\, dt < +\infty$ (see Example 2.1 (b)). We say that $\varphi \in \Phi$ satisfies the *condition* (∞), if there exists a sequence (f_k) of measurable, nonnegative functions, such that $\int_A f_k(t)\, dt < +\infty$ for every set $A \in \Sigma$ of finite measure and $k = 1, 2, \ldots$, and such that for every $u \geq 0$ the inequality

$$u \leq \frac{1}{k}\varphi(t, u) + f_k(t)$$

holds for almost all $t \in \Omega$. It is easily seen that if $\varphi \in \Phi$ satisfies (∞), then for every set $A \in \Sigma$ of finite measure and every measurable function f on Ω such that $f\chi_A \in L^\varphi(\Omega)$ there holds $f\chi_A \in L^1(\Omega)$. Moreover, we define the integral means (Steklov functions) f_r of a locally integrable function f on Ω as follows. Let $\widetilde{f}(s) = f(s)$ for $s \in \Omega$, $\widetilde{f}(s) = 0$ for $s \in \mathbb{R}^n \setminus \Omega$ and let $B_r(t)$ be the closed ball in $\mathbb{R}^n$ with centre at the point $t \in \Omega$ and with radius $r > 0$; by m_r we denote the volume of $B_r(t)$. The *integral means* f_r of f are defined by the formula

$$f_r(t) = \frac{1}{m_r} \int_{B_r(t)} \widetilde{f}(s)\, ds$$

for $t \in \Omega$ and $r > 0$.

Let $E^\varphi(\Omega)$ denote the space of all finite elements $f \in L^\varphi(\Omega)$, i.e. all functions $f \in L^\varphi(\Omega)$ such that $\rho(\lambda f) < +\infty$ for every $\lambda > 0$. Both spaces $L^\varphi(\Omega)$ and $E^\varphi(\Omega)$ are Banach spaces with respect to the norm $\|\cdot\|_\rho$, generated by ρ. There holds the following theorem ([153], Theorem 9.12).

Theorem 7.7. *Let a function $\varphi \in \Phi$ be locally integrable and let it satisfy the condition (∞). Let $\mathcal{A} \subset E_\rho^0(\Omega)$.*

1. *Suppose the following conditions hold:*

   (a) *there exists an $M > 0$ such that $\|f\|_\rho \leq M$ for all $f \in \mathcal{A}$,*

   (b) *for every $\varepsilon > 0$ there exists a compact set $A \subset \Omega$ such that for every $f \in \mathcal{A}$ there holds the inequality $\|f\chi_{\Omega\setminus A}\|_\rho < \varepsilon$,*

   (c) *for every $\varepsilon > 0$ and every compact set $A \subset \Omega$ there exists a number $r_0 > 0$ such that for every $f \in \mathcal{A}$ and every r satisfying the inequalities $0 < r < r_0$, there holds the inequality $\|(f - f_r)\chi_A\|_\rho < \varepsilon$.*

   Then $\mathcal{A}$ is conditionally compact in $E^\varphi(\Omega)$.

2. *Let us additionally assume that φ satisfies the following condition: there exist numbers $\delta > 0$, $C \geq 1$, a set $A \in \Sigma$, $A \subset \Omega$ of measure zero and a function*

$h : \Omega \times \Omega \to \mathbb{R}_0^+$ integrable with respect to the second variable and satisfying the condition $\sup_{|s| \le \delta} \int_\Omega h(s,t)\, dt < +\infty$, such that for every $u \ge 0$, $|s| \le \delta$ and $t \in \Omega \cap (\Omega + s) \setminus A$ there holds the inequality

$$\varphi(t - s, u) \le \varphi(t, Cu) + h(s, t).$$

Then if $\mathcal{A} \subset E^\varphi(\Omega)$ is conditionally compact, then the conditions (a), (b) *and* (c) *are satisfied.*

Remark 7.2. Concerning the condition in 2., compare with Example 1.10 (b). This condition is trivially satisfied if the function $\varphi(t, u)$ is independent of the variable t.

7.4 Bibliographical notes

The problem of the domain Dom T of the operator T was investigated in [161], [147] and [23], where one can find versions of Theorems 7.2 and 7.3. Theorems 7.5 and 7.6 were given in [24]. Theorem 7.7 is due to A. Kamińska [123] and to A. Kamińska and R. Płuciennik [124]. As regards Remark 7.1, a detailed exposition may be found in [195], (see also [211] and [193]). For a classical theory of nonlinear integral equations see e.g. [132], while a recent exposition can be found in [113]. Recent results about complete continuity of Urysohn integral operators and applications to integral equations can be found in [193], [194], [195]. Other extensions of the theory of integral equations can be found in [2], in which multivalued operators are considered, and in [4].

Chapter 8

Uniform approximation by sampling type operators. Applications in signal analysis

In this chapter we will consider the problem of approximating a function f, belonging to a certain functional space, by means of a general family of nonlinear integral operators. The main idea is that of building up this family in such a way that it can reproduce, in particular cases, several classical families of nonlinear integral operators of approximation theory. To this end, as in Chapter 4, we will consider functions acting from a Hausdorff locally compact topological group to $\mathbb{R}$. But one of the main interests of this chapter is the fact that among the families of operators we will deal with, there are the so-called *generalized sampling operators* in their nonlinear form. The family of linear generalized sampling operators or sampling series has been introduced and studied by P. L. Butzer and his school in Aachen. The study of its nonlinear form with respect to approximation and to rates of convergence gives applications in signal processing.

8.1 Classical results

In the last century, Whittaker, Kotel'nikov and Shannon formulated the famous WKS-sampling theorem, which says that given a function $f \in L^2(\mathbb{R}) \cap C^0(\mathbb{R})$ (being $C^0(\mathbb{R})$ the space of all continuous functions on $\mathbb{R}$),which has the support of its Fourier transform $\hat{f}$ contained in an interval $[-\pi w, \pi w]$, for $w > 0$, it is possible to reconstruct f on the whole real time-axis from the sequences $f\left(\frac{k}{w}\right)$ of its sampled values, by means of the interpolation series (see Figure 8.1)

$$f(t) = \sum_{k=-\infty}^{+\infty} f\left(\frac{k}{w}\right) \operatorname{sinc}[\pi(wt - k)], \quad t \in \mathbb{R}. \tag{8.1}$$

Here $\operatorname{sinc}(t) = \sin t/t$ for $t \neq 0$ and $\operatorname{sinc}(0) = 1$. Such interpolation takes into account the behaviour of the function f only at its sampled values $f\left(\frac{k}{w}\right)$ computed just at the "nodes" $\frac{k}{w}$, for $k \in \mathbb{Z}$, uniformly spaced on the whole real axis. The interpolation (8.1) is "free" in the sense that the sequence $\left\{f\left(\frac{k}{w}\right)\right\}_{k\in\mathbb{Z}}$, which belongs to l^2, can be an arbitrary sequence $\{c_k\}_{k\in\mathbb{Z}} \in l^2$. Moreover, the Parseval identity holds,

$$\frac{1}{w}\sum_{k=-\infty}^{+\infty}\left|f\left(\frac{k}{w}\right)\right|^2 = \int_{\mathbb{R}} |f(t)|^2\,dt.$$

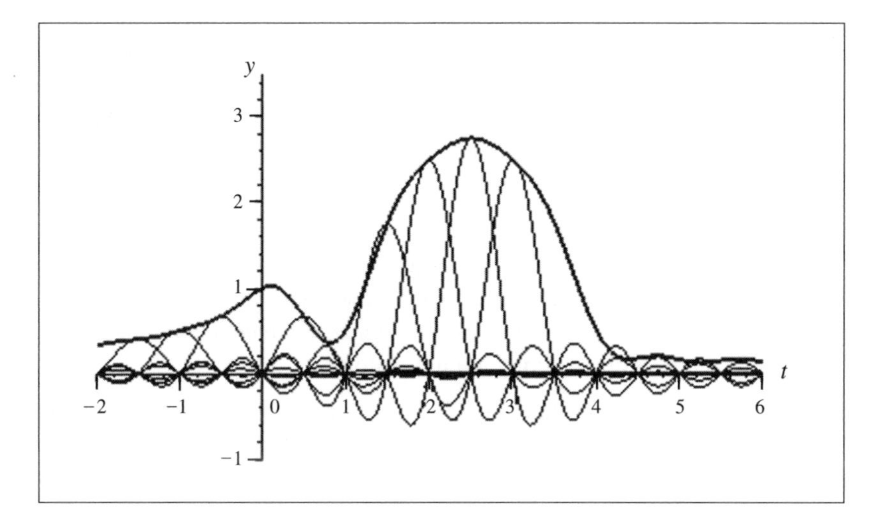

Figure 8.1. Interpolation by sampling

Now taking into account that, in the language of signal theory, a function f belonging to $L^2(\mathbb{R})$ is regarded as a signal which has *finite energy* and that the support of its Fourier transform denotes the *spectrum of f*, the WKS-sampling theorem can be formulated in the language of transmission information theory as follows (see [46]).

Let f be a signal of finite energy on $\mathbb{R}$ with bounded frequency spectrum contained in $[-\pi w, \pi w]$, which means that this signal does not contain frequencies higher than $\frac{w}{2}$ cycles for second. Moreover, let the signal have a certain communication channel. In order to recover this signal at the output of this communication channel it is sufficient to transmit over this channel only the values $f\left(\frac{k}{w}\right)$ of the signal at the nodes $\frac{k}{w}$.

Here $\Delta t = \frac{1}{w}$ denotes the *time "interval" between samples*, $\frac{\pi w}{2\pi} = \frac{w}{2}$ denotes the *bandwith of f* which is measured in cycles per unit time, while $R = \frac{1}{\Delta t}$ denotes the *sampling rate* measured in samples per unit time. H. Nyquist has marked out the meaning of the "interval" (number) $\frac{1}{w}$ for telegraphy and C. Shannon called this "interval" the *Nyquist interval corresponding to the frequency* $[-\pi w, \pi w]$. Then $R = w$ is called the *Nyquist rate* which represents the theoretical minimum sampling rate in order to reconstruct the signal completely.

Of course it is also possible to sample signals having a sampling rate bigger than R, which means the use of a thinner sequence of sample values, and this is the case of so-called *oversampling*. In practice, oversampling occurs very often since a "fudge factor" must be introduced due to the fact that sampling and interpolation cannot exactly match the theoretical one; this is the case of the so called *round-off error* or *quantization error*. Moreover, another error in time occurs when the samples cannot be taken just at the instants $\frac{k}{w}$; this error is the so-called *time jitter error*. A concrete example of an oversampling phenomenon is the compact disk player; indeed, taking into account that the highest audible frequency is approximately 18.000 cycles per second (Hertz), depending on the listener, according to the previous theory an audio signal must be sampled at least 36.000 times per second in order that the signal be reconstructed completely; but the actual rate on compact disk players is usually 44.000 samples per second.

Sometimes it also happens that one does not have at own disposal a reasonable number of samples in order to reconstruct the signal completely. This happens when the distance between the nodes is greater than the Nyquist "interval" and we have the so-called *undersampling*. In this case the *aliasing phenomenon* appears and the spectral replics of the sampled signal are not disjoint.

Just to point out the importance of the WKS-sampling theorem in the applications, we remark that besides communication theory, there are several applications to medicine through the use of image processing; only as an example, magnetic resonance imaging (MRI) can be considered .

But in practice, the interpolation formula (8.1) has some disadvantages. First of all, according to (8.1), in order to reconstruct the signal completely, the number of sample values should be infinite, which in practice does not occur. Furthermore, if we fix an instant t_0 as the present time, then formula (8.1) means that one should know the samples of the signal not only in the past of the instant t_0, but also in the future, that is for $\frac{k}{w} > t_0$. Still more, in the WKS-sampling theorem, the signal should be band-limited, which is a rather restrictive assumption; indeed if $f \in L^2(\mathbb{R})$ is a band-limited function, by Paley–Wiener theorem, this implies that f is the restriction to the real axis of an entire function of exponential type πw, which means that the function is extremely smooth; moreover, such a function cannot be simultaneously duration limited, and in practice most of the signals have the last property.

In order to avoid the above disadvantages, P. L. Butzer and his school replaced the *sinc* function in formula (8.1) by a function φ which is continuous with compact support contained in a real interval, obtaining an *approximate* sampling formula. Clearly, by using such a function φ one only needs to know a finite number of sample values and, if the interval containing the support belongs to $\mathbb{R}^+$, then the sample values can be taken only from the past, which means one is dealing with prediction of the signal. Moreover, in this case the signal should not be necessarily band-limited. Of course, one cannot expect to obtain an interpolation formula like the above one in (8.1), but we will need approximation results in order to reconstruct the signal f.

Namely, we consider a family of discrete operators, called *generalized sampling operators* of the form

$$(S_w^\varphi f)(t) = \sum_{k=-\infty}^{+\infty} f\left(\frac{k}{w}\right)\varphi(wt-k), \quad t \in \mathbb{R},\ k \in \mathbb{Z},\ w > 0, \tag{8.2}$$

and we will establish pointwise and uniform convergence of $S_w^\varphi f$ toward f, as $w \to +\infty$, together with some results concerning the rate of approximation of $(S_w^\varphi f - f)$.

In the following we will denote by $C(\mathbb{R})$ the space of all uniformly continuous and bounded functions $f : \mathbb{R} \to \mathbb{R}$, endowed with the norm $\|f\|_\infty = \sup_{t\in\mathbb{R}} |f(t)|$, and by $C^{(r)}(\mathbb{R})$ the space of all functions $f \in C(\mathbb{R})$ such that there exists the r-th derivative, $r \in \mathbb{N}$, and $f^{(r)} \in C(\mathbb{R})$. Finally, we will denote by $C_c(\mathbb{R})$ and by $C_c^{(r)}(\mathbb{R})$, $r \in \mathbb{N}$, the subspaces of $C(\mathbb{R})$ and of $C^{(r)}(\mathbb{R})$ consisting of functions with compact support.

Now, for $\varphi \in C_c(\mathbb{R})$ and $f \in C(\mathbb{R})$, we consider the series (8.2).

Since φ has compact support, for every fixed w the series (8.2) consists of only a finite number of non-zero terms, i.e., of those $k \in \mathbb{Z}$ for which $wt - k$ belongs to the support of φ. Moreover, it is easy to observe that $S_w^\varphi : C(\mathbb{R}) \to C(\mathbb{R})$, and that the following estimate can be established

$$\|S_w^\varphi\|_\infty \leq m_0(\varphi)\|f\|_\infty,$$

where $m_0(\varphi) = \sup_{u\in\mathbb{R}} \sum_{k=-\infty}^{+\infty} |\varphi(u-k)| < +\infty$.

The following theorem holds.

Theorem 8.1. *Let $\varphi \in C_c(\mathbb{R})$ be such that*

$$\sum_{k=-\infty}^{+\infty} \varphi(u-k) = 1, \quad u \in \mathbb{R}. \tag{8.3}$$

If $f : \mathbb{R} \to \mathbb{R}$ is a continuous function at $t_0 \in \mathbb{R}$, then

$$\lim_{w\to+\infty} (S_w^\varphi f)(t_0) = f(t_0). \tag{8.4}$$

Moreover, if $f \in C(\mathbb{R})$, then

$$\lim_{w\to+\infty} \|S_w^\varphi f - f\|_\infty = 0. \tag{8.5}$$

Proof. First we prove (8.4). Given $\varepsilon > 0$, by the continuity of f at t_0, there exists $\delta > 0$ such that

$$|f(t_0) - f(k/w)| < \varepsilon$$

for $|t_0 - k/w| \leq \delta$. Now, if $w > 0$, by (8.3) we may write

$$\begin{aligned}|f(t_0) - (S_w^\varphi f)(t_0)| &= \Big| \sum_{k=-\infty}^{+\infty} f(t_0)\varphi(wt_0 - k) - \sum_{k=-\infty}^{+\infty} f(k/w)\varphi(wt_0 - k) \Big| \\ &\leq \sum_{k=-\infty}^{+\infty} |f(t_0) - f(k/w)||\varphi(wt_0 - k)| \\ &= \Big(\sum_{(1)} + \sum_{(2)}\Big) |f(t_0) - f(k/w)||\varphi(wt_0 - k)| \\ &= I_1 + I_2,\end{aligned}$$

where $\sum_{(1)}$ is the sum over those $k \in \mathbb{Z}$ for which $|wt_0 - k| \leq \delta w$, while $\sum_{(2)}$ is that over those $k \in \mathbb{Z}$ such that $|wt_0 - k| > \delta w$.

Now $I_1 < \varepsilon \sum_{k=-\infty}^{+\infty} |\varphi(wt_0 - k)| \leq \varepsilon\, m_0(\varphi)$. Moreover, for $\delta > 0$ we may choose $w > 0$ so large that the support of φ is contained in $[-\delta w, \delta w]$. Therefore $I_2 = 0$, and (8.4) follows.

The uniform convergence is proved analogously because, by uniform continuity of f we can choose δ independent of $t \in \mathbb{R}$. □

The following corollary shows that taking the function φ with compact support in $\mathbb{R}^+$, (8.4) gives a prediction of the signal f.

Corollary 8.1. *Suppose that the assumptions of Theorem* 8.1 *are satisfied. Then if the function φ has compact support in $\mathbb{R}^+$, then at every point t_0 of continuity of f, we have*

$$\lim_{w\to+\infty} (S_w^\varphi f)(t_0) = \lim_{w\to+\infty} \sum_{(k/w)<t_0} f(k/w)\varphi(wt_0 - k) = f(t_0). \tag{8.6}$$

Proof. The proof is an easy consequence of the fact that $\varphi(wt_0 - k) = 0$ if $k/w \geq t_0$. Hence

$$(S_w^\varphi f)(t_0) = \sum_{k/w<t_0} f(k/w)\varphi(wt_0 - k),$$

and the assertion follows by Theorem 8.1. □

Remark 8.1. (a) We point out that (8.6) gives the prediction of the signal at an instant t_0 by the knowledge of only a finite number of sample values chosen from the past of t_0.

(b) It is important to observe that condition (8.3) is not only sufficient in order to obtain the convergence result (8.4); indeed, putting $f(t) = 1$, it is easy to see that it becomes necessary too.

(c) Results concerning the approximation of functions having jump discontinuities, by sampling sums can be found in [71]; there it is proved that at points of discontinuity, pointwise approximation is not possible in general.

According to the above Remark 8.1 (b), it is important to have conditions on the function φ such that (8.3) holds; indeed, in general, it is not easy to decide whether a function $\varphi \in C_c(\mathbb{R})$ satisfies (8.3) or not. In order to do this, the following theorem will be helpful.

Theorem 8.2. *If $\varphi \in C_c(\mathbb{R})$, then condition* (8.3) *is equivalent to:*

$$\sqrt{2\pi}\,\hat{\varphi}(2\pi k) = \begin{cases} 1, & \text{if } k = 0 \\ 0, & \text{if } k \in \mathbb{Z} \setminus \{0\}. \end{cases} \tag{8.7}$$

The interested reader can find a proof in [79]; it is a consequence of Poisson's summation formula (see [67])

$$\frac{1}{\sqrt{2\pi}} \sum_{k=-\infty}^{+\infty} \varphi(u-k) \cong \sum_{k=-\infty}^{+\infty} \hat{\varphi}(2\pi k) e^{i2\pi k u},$$

where $\cong$ means that the second series is the Fourier series of the 1-periodic function on the left.

By means of the above Theorem 8.2, it is possible to furnish examples of functions $\varphi \in C_c(\mathbb{R})$ satisfying (8.7), and hence equivalently (8.3).

Example 8.1. For $n \in \mathbb{N}$, we define the *central B-spline* of order n by

$$M_n(t) = \frac{1}{(n-1)!} \sum_{j=0}^{n} (-1)^j \binom{n}{j} \left(\frac{n}{2} + t - j\right)_+^{n-1},$$

where, $x_+^r = \begin{cases} x^r, & x \geq 0 \\ 0, & x < 0 \end{cases}$ for $x \in \mathbb{R}$, $r \in \mathbb{N}$.

For $n = 2$, we obtain the *roof-function* (see Figure 8.2),

$$M_2(t) = \begin{cases} 1 - |t|, & |t| \leq 1 \\ 0, & |t| > 1 \end{cases}$$

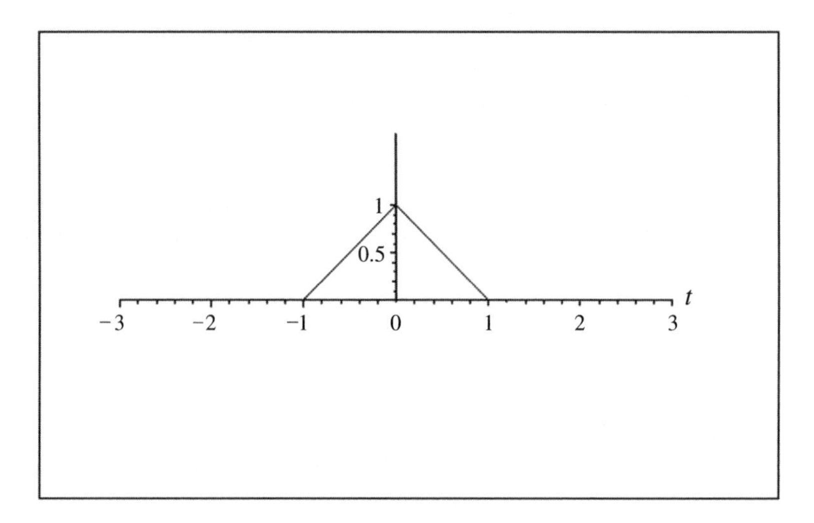

Figure 8.2. The B-spline M_2

while for $n = 3$ we get (see Figure 8.3),

$$M_3(t) = \begin{cases} \frac{1}{2}\left(|t| + \frac{3}{2}\right)^2 - \frac{3}{2}\left(|t| + \frac{1}{2}\right)^2, & |t| \le \frac{1}{2} \\ \frac{1}{2}\left(-|t| + \frac{3}{2}\right)^2, & \frac{1}{2} < |t| \le \frac{3}{2} \\ 0, & |t| > \frac{3}{2}. \end{cases}$$

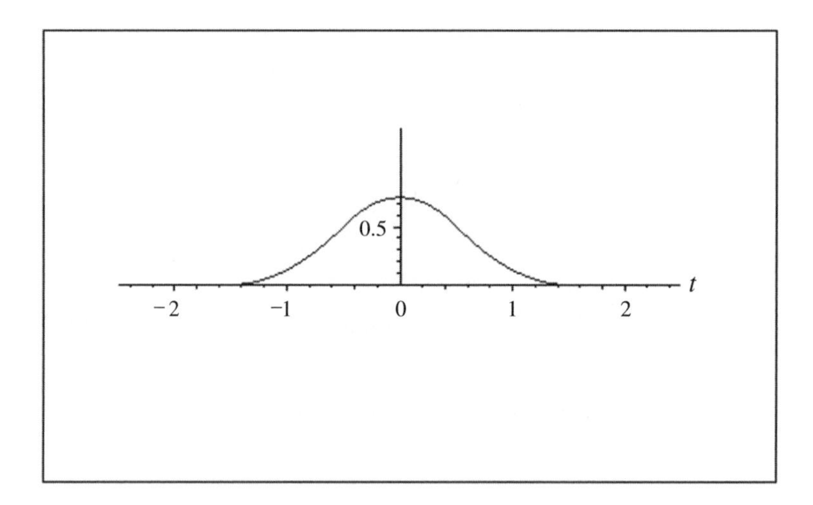

Figure 8.3. The B-spline M_3

For $n \geq 3$ it is convenient to use the following recursive formula (see e.g. [176])

$$M_n(t) = \frac{((n/2)+t)M_{n-1}(t+1/2) + ((n/2)-t)M_{n-1}(t-1/2)}{n-1}.$$

Moreover, for the Fourier transform we have

$$\sqrt{2\pi}\,\widehat{M_n}(\lambda) = \left[\frac{\sin(\lambda/2)}{\lambda/2}\right]^n, \quad \lambda \in \mathbb{R},\ n \in \mathbb{N}, \tag{8.8}$$

and hence in particular (8.8) gives

$$\widehat{M_n}(2\pi k) = 0, \quad k \in \mathbb{Z} \setminus \{0\}, \qquad \widehat{M_n}(0) = \frac{1}{\sqrt{2\pi}}.$$

Now in order to study the rate of approximation in (8.5), it is necessary to assume some further conditions besides (8.3). To this aim, for $r \in \mathbb{N} \cup \{0\}$ and $\varphi \in C_c(\mathbb{R})$, we put

$$m_r(\varphi) = \sup_{u \in \mathbb{R}} \sum_{k=-\infty}^{+\infty} |u-k|^r |\varphi(u-k)|.$$

Now the following theorem on the order of approximation can be formulated.

Theorem 8.3. *Let $\varphi \in C_c(\mathbb{R})$ and suppose that for some $r \in \mathbb{N} \setminus \{1\}$, there holds*

$$\sum_{k=-\infty}^{+\infty} (u-k)^j \varphi(u-k) = \begin{cases} 1, & j=0, \\ 0, & j=1,2,\dots,r-1 \end{cases} \tag{8.9}$$

for every $u \in \mathbb{R}$. Then

$$\|f - S_w^\varphi f\|_\infty \leq \frac{m_r(\varphi)}{r!}\|f^{(r)}\|_\infty w^{-r}, \tag{8.10}$$

for $f \in C^{(r)}(\mathbb{R})$ and $w > 0$.

Proof. Applying to the function f the Taylor expansion formula in the integral form with order r, we may write

$$f(u) = \sum_{h=0}^{r-1} \frac{f^{(h)}(t)}{h!}(u-t)^h + \frac{1}{(r-1)!}\int_t^u f^{(r)}(y)(u-y)^{r-1}\,dy.$$

Hence

$$
\begin{aligned}
&(S_w^{\varphi} f)(t) - f(t) \\
&\quad = \sum_{k=-\infty}^{+\infty} f(k/w)\varphi(wt-k) - f(t) \\
&\quad = \sum_{k=-\infty}^{+\infty} \sum_{h=0}^{r-1} \frac{f^{(h)}(t)}{h!}((k/w)-t)^h \varphi(wt-k) \\
&\qquad + \sum_{k=-\infty}^{+\infty} \frac{1}{(r-1)!} \left\{ \int_t^{k/w} f^{(r)}(y)((k/w)-y)^{r-1}\,dy \right\} \varphi(wt-k) - f(t)
\end{aligned}
$$

for every $t \in \mathbb{R}$. Taking into account (8.9), we have

$$
\begin{aligned}
&(S_w^{\varphi} f)(t) - f(t) \\
&\quad = \sum_{h=1}^{r-1} \frac{(-1)^h}{w^h} \frac{f^{(h)}(t)}{h!} \sum_{k=-\infty}^{+\infty} (wt-k)^h \varphi(wt-k) \\
&\qquad + \sum_{k=-\infty}^{+\infty} \frac{1}{(r-1)!} \left\{ \int_t^{k/w} f^{(r)}(y)((k/w)-y)^{r-1}\,dy \right\} \varphi(wt-k) \\
&\quad = \sum_{k=-\infty}^{+\infty} \frac{1}{(r-1)!} \left\{ \int_t^{k/w} f^{(r)}(y)((k/w)-y)^{r-1}\,dy \right\} \varphi(wt-k).
\end{aligned}
$$

Estimating now the above integral via

$$
\left| \int_t^{k/w} f^{(r)}(y)((k/w)-y)^{r-1}\,dy \right| \le \|f^{(r)}\|_\infty \frac{w^{-r}}{r} |k-wt|^r,
$$

the assertion follows from the definition of $m_r(\varphi)$. □

Remark 8.2. We remark that, if $r = 1$, then (8.9) reduces to (8.3). This is important since it means that the above estimate (8.10) for $r = 1$ holds under the assumptions of Theorem 8.1.

According to Theorem 8.2, there follows an analogous condition equivalent to (8.9).

Theorem 8.4. *If $\varphi \in C_c(\mathbb{R})$, then condition* (8.9) *is equivalent to*

$$
\hat{\varphi}^{(j)}(2\pi k) = \begin{cases} 1/\sqrt{2\pi}, & k = j = 0 \\ 0, & k \in \mathbb{Z} \setminus \{0\},\ j = 0 \\ 0, & k \in \mathbb{Z},\ j = 1, 2, \dots, r-1,\ r \in \mathbb{N},\ r > 1. \end{cases} \tag{8.11}
$$

Proof. The proof is again a consequence of Poisson's summation formula applied to the function $(-iu)^j\varphi(u)$. Then by a well-known property of the derivative of order r of the Fourier transform, we obtain

$$\frac{(-i)^j}{\sqrt{2\pi}}\sum_{k=-\infty}^{+\infty}(u-k)^j\varphi(u-k)\cong\sum_{k=-\infty}^{+\infty}\hat{\varphi}^{(j)}(2\pi k)e^{i2\pi ku},$$

and so the proof follows with reasonings similar to those of Theorem 8.2 (see also [79]). □

Example 8.2. If $r=2$, the kernel $\varphi_2(t)=3M_2(t-2)-2M_2(t-3)$, satisfies (8.11) (see Figure 8.4).

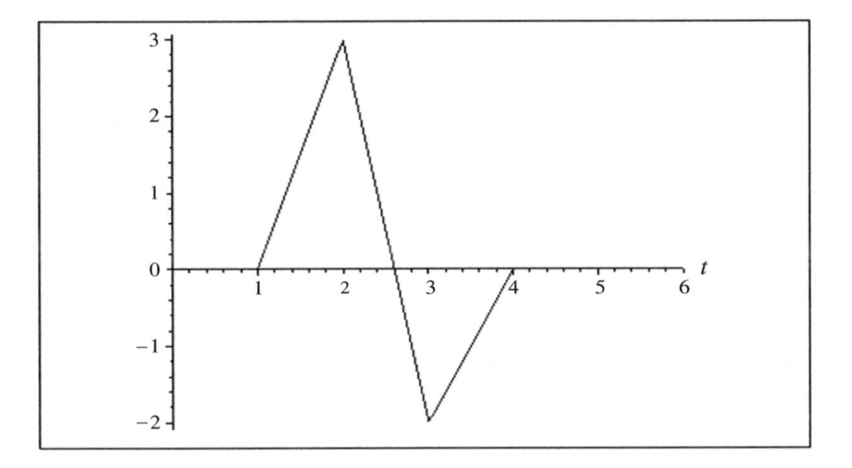

Figure 8.4. The kernel φ_2

Moreover, in this case $m_r(\varphi_2)/r!\leq 15$, and

$$\|f-S_w^{\varphi_2}f\|_\infty=O(w^{-2}),\quad w\to+\infty.$$

If $r=3$, the kernel

$$\varphi_3(t)=\frac{1}{8}\{47M_3(t-2)-62M_3(t-3)+23M_3(t-4)\}$$

satisfies (8.11) (see Figure 8.5); in this case $m_r(\varphi_3)/r!\leq 54$ and

$$\|f-S_w^{\varphi_3}f\|_\infty=O(w^{-3}),\quad w\to+\infty.$$

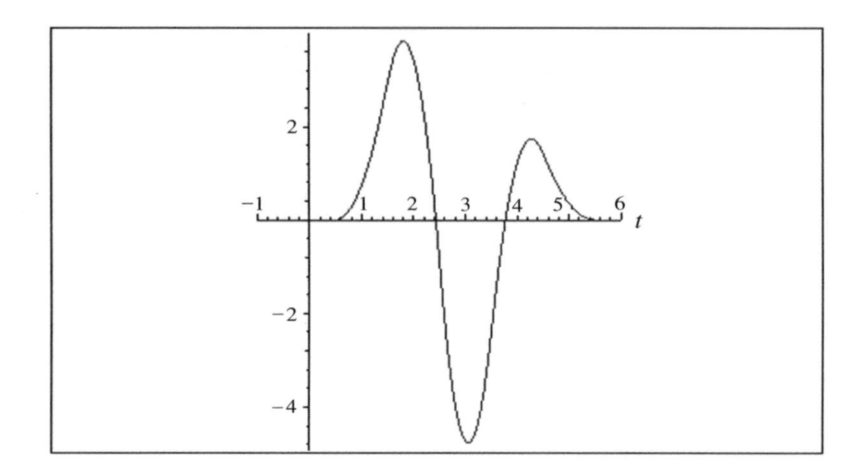

Figure 8.5. The kernel φ_3

The construction of such functions is based on the solution of linear systems in the complex plane. We remark that the above kernels are constructed in such a way that they satisfy (8.9) and that they have compact support contained in $(0, +\infty)$, just as for the case of the prediction of a signal. For example, it is possible to show that $S_w^{\varphi_2} f$ predicts a signal at least $\frac{1}{w}$ units ahead with error $\mathcal{O}(w^{-2})$, and the associate sampling series consists only of three terms, i.e., those for $k = j-3, j-2, j-1$ for which $t - \frac{4}{w} < \frac{k}{w} < t - \frac{1}{w} < t$. For readers interested in results of this type, we suggest the paper of P. L. Butzer and R. L. Stens [79].

Remark 8.3. Also inverse results concerning the order of approximation have been established. In [79] it is proved that using spline kernels it is not possible to approximate functions of the class $C^{(s)}(\mathbb{R})$ with $s \le r$ with a rate better than $\mathcal{O}(w^{-s})$. Moreover, it is also proved that for spline kernels of order r, the best possible order of approximation which can be obtained for non-constant functions f is $\mathcal{O}(w^{-r})$, even if f is arbitrarily smooth.

8.2 Uniform convergence for a class of nonlinear integral operators

In this section we will construct a general family of nonlinear integral operators which contains, in particular, a nonlinear version of the generalized sampling operator introduced in the previous section. For such nonlinear generalized sampling operators, we will study uniform approximation results. From a mathematical point of view, the theory developed gives a unitary approach to the study of convergence results for several families of nonlinear integral operators, very common in approximation theory (see e.g. Chapter 3). Moreover, from the point of view of applications to signal processing, the study of nonlinear approximation processes is important since it may describe nonlinear systems in which the signal computed, during its filtering, generates new frequencies.

As in Chapter 4, Ω will be a locally compact (and σ-compact) topological group G, and H will denote another locally compact topological group with its Haar measure μ_H on the class of Borel sets $\mathcal{B}(H)$. We suppose G abelian, but unimodularity would be sufficient for our theory. Moreover, we will denote by $\mathcal{U}$ a base of symmetric neighbourhoods of the neutral element $\theta \in G$, and by local compactness we can use a base with (measurable) symmetric compact neighbourhoods of θ. Let $\{h_w\}_{w \in W}$ be a net of functions (here W is a set of indices), $h_w : H \to G$ such that h_w is a homeomorphism between H and $h_w(H)$.

Let $\mathcal{X} = L^0(G)$ denote the space of all Borel measurable real-valued functions defined on G.

As concerns convergence, in this section we will use, concerning the set W, the same notations as in Section 3.2., assuming that $w_0 = +\infty$; i.e., we will assume that W is an unbounded subset of the interval $]0, +\infty[$, $+\infty$ being an accumulation point of the set W. This is achieved by taking into account the nature of the classical results given in Section 8.1, where $w_0 = +\infty$.

For every $U \in \mathcal{U}$, $w \in W$ and $s \in G$ let us now define the sets

$$U_{s,w} = \{t \in H : s - h_w(t) \in U\} = h_w^{-1}(s + U).$$

Moreover, let $\{L_w\}_{w \in W}$ be a family of measurable functions $L_w : G \to \mathbb{R}$, with $L_w(s - h_w(\cdot)) \in L^1_{\mu_H}(H)$ for almost all $s \in G$.

We will suppose that for every $w \in W$, $L_w : G \to \mathbb{R}^+$ is a measurable function which satisfies the following assumptions:

$(L_w.1)$ for every $U \in \mathcal{U}$,

$$\lim_{w \to +\infty} \int_{H \setminus h_w^{-1}(s+U)} L_w(s - h_w(t))\, d\mu_H(t) = 0,$$

uniformly with respect to $s \in G$,

$(L_w.2)$ there is a constant $N > 0$ such that, for every $s \in G$ and $w \in W$,

$$\int_H L_w(s - h_w(t))\, d\mu_H(t) \leq N.$$

If the family $\{L_w\}_{w \in W}$ satisfies $(L_w.1)$ and $(L_w.2)$ then we will write $\{L_w\}_{w \in W} \subset \mathcal{L}_w$.

Later on, we will show that in some particular cases, assumption $(L_w.2)$ implies $(L_w.1)$.

Now we introduce an important class of kernels.

Let $\{K_w\}_{w\in W}$ be a net of globally measurable functions $K_w : G \times \mathbb{R} \to \mathbb{R}$ such that $K_w(s - h_w(\cdot), u) \in L^1_{\mu_H}(H)$ for every $u \in \mathbb{R}$, $s \in G$, and there hold:

$(K_w.1)$ $K_w(s, 0) = 0$ for $s \in G$,

$(K_w.2)$ the family $\{K_w\}_{w\in W}$ is an (L_w, ψ)-Lipschitz kernel, i.e., there exists a family $\{L_w\}_{w\in W} \subset \mathcal{L}_w$ such that

$$|K_w(s, u) - K_w(s, v)| \leq L_w(s)\psi(|u - v|)$$

for $s \in G, u, v \in \mathbb{R}$ and $\psi : \mathbb{R}_0^+ \to \mathbb{R}_0^+$ is a function belonging to the class Ψ (compare with the definitions given in Sections 3.1 and 4.1, dealing with families of functions and ψ being independent of the parameter $t \in \Omega = G$),

$(K_w.3)$ for every $n \in \mathbb{N}$ and $w \in W$, putting

$$r_n^w(s) = \sup_{\frac{1}{n}\leq|u|\leq n} \left| \frac{1}{u} \int_H K_w(s - h_w(t), u)\, d\mu_H(t) - 1 \right|,$$

we have $\lim_{w\to+\infty} r_n^w(s) = 0$, uniformly with respect to $s \in G$.

From now on, if a family of kernels $\{K_w\}_{w\in W}$ satisfies $(K_w.i)$, $i = 1, 2, 3$, then we will write $\{K_w\}_{w\in W} \subset \mathcal{K}_w$, and if the function $\psi \in \Psi$ is also concave, we will write $\psi \in \widetilde{\Psi}$.

Throughout this chapter, we will deal with the following net of nonlinear integral operators

$$(T_w f)(s) = \int_H K_w(s - h_w(t), f(h_w(t)))\, d\mu_H(t) \tag{8.12}$$

defined for every $f \in \operatorname{Dom} \mathbb{T} = \bigcap_{w\in W} \operatorname{Dom} T_w$, i.e., for every $f \in L^0(G)$ for which $(T_w f)(s)$ exists, as an Haar integral, for every $s \in G$ and for every $w \in W$.

In order to prove the main approximation result, we need the following preliminary lemma.

Lemma 8.1. *If $\{K_w\}_{w\in W} \subset \mathcal{K}_w$, then there exist $\overline{w} \in W$ and $r > 0$ such that for every $w \in W$, $w > \overline{w}$, and every $s \in G$, we have*

$$\int_H L_w(s - h_w(t))\, d\mu_H(t) > r.$$

Proof. Let $n \in \mathbb{N}$ be fixed. From the singularity, there is a $\overline{w} \in W$ such that, for every $w \in W$ with $w > \overline{w}$ and for every $s \in G$,

$$\sup_{\frac{1}{n}\leq|u|\leq n} \left| \frac{1}{u} \int_H K_w(s - h_w(t), u)\, d\mu_H(t) - 1 \right| < \frac{1}{2}$$

or, for every $w \in W$, $w > \overline{w}$, $s \in G$ and $\frac{1}{n} \leq |u| \leq n$,

$$\left| \frac{1}{u} \int_H K_w(s - h_w(t), u)\, d\mu_H(t) - 1 \right| < \frac{1}{2}.$$

This implies the majorization

$$\frac{1}{u} \int_H K_w(s - h_w(t), u)\, d\mu_H(t) > \frac{1}{2} \tag{8.13}$$

for every $w \in W$, $w > \overline{w}$, $s \in G$ and $\frac{1}{n} \leq |u| \leq n$.

Suppose now that the assertion of this lemma is false, i.e., for every $w \in W$ and for every $r > 0$,

$$\int_H L_{w'}(s - h_{w'}(t))\, d\mu_H(t) \leq r, \tag{8.14}$$

for some $s \in G$ and for some $w' \in W$, $w' > w$.

In particular we have that, for every $r > 0$, (8.14) holds for some $s \in G$ and for some $w' \in W$, $w' > \overline{w}$.

Now, for such $w' \in W$ and $s \in G$ and for a fixed $n \in \mathbb{N}$, we may write, for every $u \in \mathbb{R}$, $\frac{1}{n} \leq |u| \leq n$,

$$\begin{aligned}
&\left| \frac{1}{u} \int_H K_{w'}(s - h_{w'}(t), u)\, d\mu_H(t) \right| \\
&\quad \leq \frac{1}{|u|} \int_H |K_{w'}(s - h_{w'}(t), u)|\, d\mu_H(t) \\
&\quad \leq \frac{1}{|u|} \int_H L_{w'}(s - h_{w'}(t)) \psi(|u|)\, d\mu_H(t) \\
&\quad = \frac{\psi(|u|)}{|u|} \int_H L_{w'}(s - h_{w'}(t))\, d\mu_H(t) \\
&\quad \leq \frac{\psi(|u|)}{|u|} r \leq \sup_{\frac{1}{n} \leq |u| \leq n} \frac{\psi(|u|)}{|u|} r = \frac{\psi(n)}{\frac{1}{n}} r \\
&\quad = n\psi(n) r.
\end{aligned}$$

Now we can choose $r > 0$ such that $n\psi(n)\, r < \frac{1}{2}$ and so we obtain a contradiction to (8.13). Hence the assertion follows. □

As in Section 8.1, let $C(G)$ be the set of all bounded and uniformly continuous functions $f : G \to \mathbb{R}$. Now we may state the following approximation theorem.

Theorem 8.5. *Let $f : G \to \mathbb{R}$, $f \in C(G)$ and suppose that $\{K_w\}_{w \in W} \subset \mathcal{K}_w$ and $\{L_w\}_{w \in W} \subset \mathcal{L}_w$. Then*

$$\|T_w f - f\|_\infty \to 0 \quad \textit{as } w \to +\infty.$$

Moreover, $T_w : C(G) \to L^\infty(G)$ *and for some constant* $N > 0$, *we have, for every* $w \in W$,

$$\|T_w f\|_\infty \le N\psi(\|f\|_\infty).$$

Proof. First we evaluate $\|T_w f\|_\infty$. By the (L_w, ψ)-Lipschitz condition and $K_w(s, 0) = 0$, for every $s \in G$, we may write, taking into account that $f \in C(G)$,

$$\begin{aligned} |(T_w f)(s)| &= \left| \int_H K_w(s - h_w(t), f(h_w(t)))\, d\mu_H(t) \right| \\ &\le \int_H L_w(s - h_w(t))\psi(|f(h_w(t))|)\, d\mu_H(t) \\ &\le N\psi(\|f\|_\infty), \end{aligned}$$

for every $s \in G$; hence we obtain $\|T_w f\|_\infty \le N\psi(\|f\|_\infty)$, and so $T_w : C(G) \to L^\infty(G)$. We now evaluate $\|T_w f - f\|_\infty$. We have

$$\begin{aligned} |T_w f(s) - f(s)| &= \left| \int_H K_w(s - h_w(t), f(h_w(t)))\, d\mu_H(t) \right. \\ &\qquad - \int_H K_w(s - h_w(t), f(s))\, d\mu_H(t) \\ &\qquad \left. + \int_H K_w(s - h_w(t), f(s))\, d\mu_H(t) - f(s) \right| \\ &\le \int_H L_w(s - h_w(t))\psi(|f(h_w(t)) - f(s)|)\, d\mu_H(t) \\ &\qquad + \left| \int_H K_w(s - h_w(t), f(s))\, d\mu_H(t) - f(s) \right| \\ &= I_1 + I_2. \end{aligned}$$

First we consider I_1. By the uniform continuity of f, for a fixed $\varepsilon > 0$, there is a compact neighbourhood $U_\varepsilon \in \mathcal{U}$ such that $|f(s + v) - f(s)| \le \varepsilon$, for every $s \in G$ and $v \in U_\varepsilon$. Moreover, by Lemma 8.1, there are $\overline{w} \in W$ and $r > 0$ such that, for every $w \in W$ $w > \overline{w}$, $\int_H L_w(s - h_w(t))\, d\mu_H(t) > r$, for every $s \in G$. We now put

$$U^\varepsilon_{s,w} = \{t \in H : s - h_w(t) \in U_\varepsilon\} = h_w^{-1}(s + U_\varepsilon), \qquad \text{for } w \in W,\ w > \overline{w}.$$

Then if $t \in H \setminus h_w^{-1}(s + U_\varepsilon)$, $s - h_w(t) \notin U_\varepsilon$ and by property $(L_w.1)$ there exists a $\overline{\overline{w}} \in W$ such that $\int_{H\setminus h_w^{-1}(s+U_\varepsilon)} L_w(s - h_w(t))\, d\mu_H(t) < \varepsilon$ for $w \in W$, $w > \overline{\overline{w}}$, uniformly with respect to $s \in G$.

Taking $\widetilde{w} = \max\{\overline{w}, \overline{\overline{w}}\}$, we have that, for $w > W$, $w > \widetilde{w}$, $h_w^{-1}(s+U_\varepsilon) \neq \emptyset$,

$$\begin{aligned} I_1 &= \int_{h_w^{-1}(s+U_\varepsilon)} L_w(s-h_w(t))\psi(|f(h_w(t))-f(s)|)\,d\mu_H(t) \\ &\quad + \int_{H\setminus h_w^{-1}(s+U_\varepsilon)} L_w(s-h_w(t))\psi(|f(h_w(t))-f(s)|)\,d\mu_H(t) \\ &\leq \psi(\varepsilon)N + \psi(2\|f\|_\infty)\varepsilon. \end{aligned}$$

In order to estimate I_2, since f is bounded, for every $\varepsilon > 0$ there is an $n \in \mathbb{N}$ such that $|f(s)| \leq n$, for every $s \in G$ and $\frac{1}{n} < \varepsilon$. Now we fix this $n \in \mathbb{N}$, and put $A_n = \{s \in G : 0 < |f(s)| < \frac{1}{n}\}$. Then, since $K_w(s,0) = 0$, for every $s \in G$, we may write

$$\begin{aligned} I_2 &= \left| \int_H K_w(s-h_w(t), f(s))\,d\mu_H(t) - f(s) \right| \\ &= \left| \int_H K_w(s-h_w(t), f(s)\chi_{A_n}(s))\,d\mu_H(t) - f(s)\chi_{A_n}(s) \right. \\ &\quad \left. + \int_H K_w(s-h_w(t), f(s)\chi_{G\setminus A_n}(s))\,d\mu_H(t) - f(s)\chi_{G\setminus A_n}(s) \right|. \end{aligned}$$

Hence there follows

$$\begin{aligned} I_2 &\leq r_n^w(s)|f(s)| + \left| \int_H K_w(s-h_w(t), f(s)\chi_{A_n}(s))\,d\mu_H(t) - f(s)\chi_{A_n}(s) \right| \\ &= I_2^1 + I_2^2. \end{aligned}$$

Now, $I_2^1 \leq \|f\|_\infty r_n^w(s)$, and since $\psi \in \Psi$,

$$\begin{aligned} I_2^2 &\leq \int_H L_w(s-h_w(t))\psi(|f(s)\chi_{A_n}(s)|)\,d\mu_H(t) + |f(s)\chi_{A_n}(s)| \\ &\leq N\psi\left(\frac{1}{n}\right) + \frac{1}{n} \\ &< N\psi(\varepsilon) + \varepsilon. \end{aligned}$$

Finally, for $w \in W$, $w > \widetilde{w}$,

$$|T_w f(s) - f(s)| \leq N\psi(\varepsilon) + \psi(2\|f\|_\infty)\varepsilon + N\psi(\varepsilon) + \varepsilon + r_n^w(s)\|f\|_\infty$$

and so, since $r_n^w(s) \to 0$ uniformly with respect to $s \in G$, we obtain

$$\limsup_{w\to+\infty} |T_w f(s) - f(s)| \leq N\psi(\varepsilon) + \psi(2\|f\|_\infty)\varepsilon + N\psi(\varepsilon) + \varepsilon$$

uniformly with respect to $s \in G$. Hence the assertion follows, $\varepsilon > 0$ being arbitrary.□

Remark 8.4. (a) Here we observe that in the case of $f(s) = u$, $u \neq 0$ a fixed number, for every $s \in G$ it follows that if $\|T_w f - f\|_\infty \to 0$ as $w \to +\infty$, then

$$\left| \int_H K_w(s - h_w(t), u)\, d\mu_H(t) - u \right| \to 0 \quad \text{as } w \to +\infty,$$

and this implies that

$$\frac{1}{u}\left\{ \int_H K_w(s - h_w(t), u)\, d\mu_H(t) - 1 \right\} \to 0 \quad \text{as } w \to +\infty,$$

i.e., $r_n^w(s) \to 0$ as $w \to +\infty$. This means that the notion of singularity is also necessary in order to have the required approximation theorem.

(b) We remark that in the particular case of $(L_w, 1)$-Lipschitz kernel $\{K_w\}_{w\in W} \subset \mathcal{K}_w$, (strongly Lipschitz kernels), i.e.,

$$|K_w(s, u) - K_w(s, v)| \leq L_w(s)|u - v|$$

for $s \in G$, $u, v \in \mathbb{R}$, and $\{L_w\}_{w\in W} \subset \mathcal{L}_w$, from Theorem 8.5 we obtain for some constant $N > 0$ the estimate

$$\|T_w f\|_\infty \leq N\|f\|_\infty.$$

(c) As a particular case, Theorem 8.5 contains the linear case, i.e., $K_w(t, u) = \widetilde{K}_w(t)u$.

8.3 Applications to the convergence results

In this section we discuss some examples of operators (8.12) to which Theorem 8.5 can be applied.

(I) Let $G = H = (\mathbb{R}^N, +)$ and $\mu_H = dt$, the Lebesgue measure. Put $h_w(t) = t$ for every $w \in W$; then we obtain

$$(T_w f)(s) = \int_{\mathbb{R}^N} K_w(s - t, f(t))\, dt, \quad s \in \mathbb{R}^N.$$

Setting $K_w(t, u) = \check{K}_w(-t, u)$, the above operators become those of the form considered in Chapters 3 and 7.

Here it is shown that the assumptions on $\{L_w\}_{w\in W}$ and $\{K_w\}_{w\in W}$ represent natural extensions of the approximate identity to the nonlinear setting. Indeed, if $K_w(t, u) = \widetilde{K}_w(t)\, u$, $(L_w.1)$ and $(L_w.2)$ become respectively

$$\lim_{w\to+\infty} \int_{|z|>\delta} |K_w(z)|\, dz = 0, \quad \text{for every } \delta > 0,$$

and

$$\text{there is a constant } N > 0 \text{ such that } \int_{\mathbb{R}^N} |K_w(z)|\, dz \le N.$$

Moreover, the (L_w, ψ)-Lipschitz condition is always satisfied as an equality with $L_w = |K_w|$ and $\psi(u) = u$, and condition $r_n^w(s) \to 0$ as $w \to +\infty$ uniformly with respect to $s \in G$, becomes $\int_{\mathbb{R}^N} K_w(z)\, dz \to 1$ as $w \to +\infty$.

(II) An analogous application gives a nonlinear version of the Mellin convolution operator of the form

$$(T_w f)(s) = (\mathcal{M}_w f)(s) = \int_0^{+\infty} K_w(st^{-1}, f(t))t^{-1}\, dt, \quad s > 0,$$

(see also Section 4.7). In order to obtain the above operators we take $G = H = (\mathbb{R}^+, \cdot)$, $\mu_H = \int t^{-1}\, dt$, $h_w(t) = t$, for every $w \in W$. In the particular case of $K_w(t, u) = \widetilde{K}_w(t)u$ we obtain the Mellin convolution operator; it is connected with the theory of "moment type operators". The assumptions on $\{L_w\}_{w\in W}$ and $\{K_w\}_{w\in W}$ are the natural extensions to nonlinear instance of the usual assumptions used for moment type kernels.

(III) Here we consider the case of discrete operators. Such applications reproduce, in particular, the nonlinear form of the generalized sampling operators (series) introduced, in the linear case, in Section 8.1.

We put $G = (\mathbb{R}, +)$ and $H = (\mathbb{Z}, +)$ with the counting measure μ_H. Given the family of homeomorphisms $\{h_w\}_{w\in W}$, $h_w : \mathbb{Z} \to h(\mathbb{Z}) \subset \mathbb{R}$, we define the kernels $(K_w)_{w\in W}$, $K_w : \mathbb{R} \times \mathbb{R} \to \mathbb{R}$, and $\{L_w\}_{w\in W}$, $L_w : \mathbb{R} \to \mathbb{R}_0^+$ with $(K_w)_{w\in W} \subset \mathcal{K}_w$ and $\{L_w\}_{w\in W} \subset \mathcal{L}_w$.

In this case $(L_w.1)$ and $(L_w.2)$ become respectively

$$\lim_{w\to+\infty} \sum_{|s-h_w(k)|>\delta} L_w(s - h_w(k)) = 0, \quad \text{for every } \delta > 0,$$

uniformly with respect to $s \in \mathbb{R}$, and

$$\sum_{k=-\infty}^{+\infty} L_w(s - h_w(k)) \le N$$

for every $s \in \mathbb{R}$, $w \in W$.

For $\{K_w\} \subset \mathcal{K}_w$, the operators $(T_w f)$ now take on the form

$$(T_w f)(s) = \sum_{k=-\infty}^{+\infty} K_w(s - h_w(k), f(h_w(k))), \quad s \in \mathbb{R},\ w \in W.$$

Therefore, it is possible to state Theorem 8.5 for the above discrete operators.

Now we will consider a particular case, namely, we take $W = \mathbb{R}^+$, G and H as above and $h_w : \mathbb{Z} \to \mathbb{R}$, defined by $h_w(k) = \frac{k}{w}$, $w > 0$, $k \in \mathbb{Z}$; moreover, we put $K_w(z, \cdot) = K(wz, \cdot)$, $w > 0$, $z \in \mathbb{R}$ where $K : \mathbb{R} \times \mathbb{R} \to \mathbb{R}$, satisfies the (L, ψ)-Lipschitz condition, i.e.,

$$|K(s,u) - K(s,v)| \leq L(s)\psi(|u - v|)$$

for $s \in \mathbb{R}$, $u, v \in \mathbb{R}$, $\psi \in \Psi$.

So we define the *nonlinear generalized sampling operators* as

$$(T_w f)(s) = \sum_{k=-\infty}^{+\infty} K\left(ws - k, f\left(\frac{k}{w}\right)\right), \quad s \in \mathbb{R}, \ w > 0. \tag{8.15}$$

We put $L_w(z) = L(wz)$, $w > 0$, $z \in \mathbb{R}$ and $L : \mathbb{R} \to \mathbb{R}_0^+$ is a measurable function.

In this case the assumptions $(L_w.1)$ and $(L_w.2)$ become respectively

(a) $$\lim_{w \to +\infty} \sum_{|ws-k| > \delta w} L(ws - k) = 0$$

and

(b) $$\sum_{k=-\infty}^{+\infty} L(ws - k) \leq N, \quad \text{for every } s \in G \text{ and } w > 0.$$

Moreover, in this case

(c) $$r_n^w(s) = \sup_{\frac{1}{n} \leq |u| \leq n} \left| \frac{1}{u} \sum_{k=-\infty}^{+\infty} K(ws - k, u) - 1 \right| \to 0, \quad \text{as } w \to +\infty,$$

uniformly with respect to $s \in \mathbb{R}$.

In the linear case, i.e., when $K_w(t,u) = \widetilde{K}_w(t)u$, the assumption $r_n^w(s) \to 0$ as $w \to +\infty$ becomes $\sum_{k=-\infty}^{+\infty} \widetilde{K}(ws - k) \to 1$ as $w \to +\infty$, and (a) and (b) hold with $|\widetilde{K}|$ instead of L. The above assumptions are those of the theory of generalized sampling series as considered in Section 8.1. In this particular case of the sampling operator, it is not difficult to prove that $(L_w.2)$ implies $(L_w.1)$, i.e., (b) implies (a) (see [175]).

Therefore we may state the following corollary.

Corollary 8.2. *Let $f : \mathbb{R} \to \mathbb{R}$, $f \in C(\mathbb{R})$ and suppose that K is (L, ψ)-Lipschitz, with $\psi \in \Psi$ and where L satisfies assumptions* (b) *and* (c). *Then*

$$\left\| \sum_{k=-\infty}^{+\infty} K(w \cdot -k, f(\frac{k}{w})) - f(\cdot) \right\|_\infty \to 0, \quad \text{as } w \to +\infty;$$

moreover

$$\Big\| \sum_{k=-\infty}^{+\infty} K(w\cdot -k, f(\frac{k}{w}))\Big\|_\infty \leq N\psi(\|f\|_\infty),$$

N being the constant of Theorem 8.5.

Remark 8.5. (a) The previous theory also subsumes a version of the *multivariate nonlinear sampling operator*. Indeed, we may take $G = (\mathbb{R}^N, +)$, $H = (\mathbb{Z}^N, +)$ with the counting measure, and $f : \mathbb{R}^N \to \mathbb{R}$.

We then define the *multivariate nonlinear sampling operator* (*series*) as the operator of the form

$$(T_w f)(s) = \sum_{\mathbb{Z}^N} K\left(ws - k, f\left(\frac{k}{w}\right)\right)$$

where $s \in \mathbb{R}^N$, $w \in \mathbb{R}^N_+$, and $K : \mathbb{R}^N \to \mathbb{R}$. Here $w = (w_1, w_2, \ldots, w_N) \in \mathbb{R}^N_+$ and we define $w_1 \leq w_2$ if and only if $w_i^1 \leq w_i^2$, $i = 1, 2, \ldots N$. Moreover, if $s = (s_1, s_2, \ldots, s_N)$ and $k = (k_1, k_2, \ldots, k_N)$, we set $ws = (w_1 s_1, w_2 s_2, \ldots, w_N s_N)$, $\frac{k}{w} = \left(\frac{k_1}{w_1}, \frac{k_2}{w_2}, \ldots, \frac{k_N}{w_N}\right)$, and $w \to +\infty$ means that $w_i \to +\infty$ for each $i = 1, 2, \ldots, N$.

(b) The previous theory holds also for more general H. Indeed, it suffices to take $(H, \mathcal{B}(H), \mu_H)$ as a locally compact Hausdorff topological space with its Borel σ-algebra $\mathcal{B}(H)$ and μ_H a regular measure on it. Moreover, the real parameter $w > 0$ can be replaced by an abstract parameter w varying in an arbitrary filtering partially ordered set $\mathcal{W}$.

8.4 Rate of uniform approximation

In this section we study the rate of approximation of $\|T_w f - f\|_\infty$. To this aim, since f is defined on a group G, we must adapt the definition of the classical Lipschitz classes to this setting.

So, let $\omega : G \times G \to \mathbb{R}^+$ be a measurable symmetric function (i.e., $\omega(x, y) = \omega(y, x)$); we say that $f : G \to \mathbb{R}$ is *ω-Lipschitz* if there exists a constant $R > 0$ such that for every $x, y \in G$, we have

$$|f(x) - f(y)| \leq R\,\omega(x, y).$$

We will assume that ω and $\{h_w\}_{w\in W}$ are connected by the following relation.

There exist a function $\lambda : \mathbb{R}^+ \to \mathbb{R}^+$ with $\lim_{w\to+\infty} \lambda(w) = 0$ and a family of functions $\{\Omega_w\}$, $\Omega_w : G \times H \to \mathbb{R}^+$ with the following property: for every $s \in G$ and $w \in W$, there exists a $s_w \in G$ such that

$$\omega(s, h_w(t)) \leq \Omega_w(s_w, t)\lambda(w).$$

From now on, if f is ω-Lipschitz and the previous relation between ω and h_w is satisfied, we will write $f \in \mathrm{Lip}_\omega$.

In particular, for $x = s$ and $y = h_w(t)$, we obtain

$$|f(s) - f(h_w(t))| \le R\,\Omega_w(s_w, t)\,\lambda(w).$$

Moreover, we will suppose that

$$r^w(s) \equiv \sup_{n\in\mathbb{N}} r_n^w(s) = \mathcal{O}(\lambda(w)) \quad \text{as } w \to +\infty,$$

uniformly with respect to $s \in G$.

Let us set $m_\omega = \sup_{w\in W} \sup_{s\in G} \int_H L_w(s - h_w(t))\,\Omega_w(s_w, t)\,d\mu_H(t)$.

We are ready to state the following

Theorem 8.6. *Let $f : G \to \mathbb{R}$ with $f \in C(G)$, and let $\{K_w\}_{w\in W} \subset \mathcal{K}_w$, $\{L_w\}_{w\in W} \subset \mathcal{L}_w$, and $\psi \in \widetilde{\Psi}$. If $f \in \mathrm{Lip}_\omega$, $m_\omega < +\infty$ and $r^w(s) = \mathcal{O}(\lambda(w))$, as $w \to +\infty$ uniformly with respect to $s \in G$, λ being the function in the definition of the class Lip_ω, then*

$$\|T_w f - f\|_\infty = \mathcal{O}(\psi(\lambda(w))) \quad \textit{as } w \to +\infty.$$

Proof. Arguing as in Theorem 8.5, we may estimate $|T_w f(s) - f(s)|$ as follows,

$$\begin{aligned}|T_w f(s) - f(s)| &\le \int_H L_w(s - h_w(t))\psi(|f(h_w(t)) - f(s)|)\,d\mu_H(t) \\ &\quad + \left|\int_H K_w(s - h_w(t), f(s))\,d\mu_H(t) - f(s)\right| \\ &= I_1(w) + I_2(w).\end{aligned}$$

Now, $I_2(w) \le r^w(s)|f(s)| \le r^w(s)\|f\|_\infty$, and hence, by the assumptions, we have $I_2(w) = \mathcal{O}(\lambda(w))$ as $w \to +\infty$. Now, since $\psi \in \widetilde{\Psi}$ is concave, then $\lambda(w) = \mathcal{O}(\psi(\lambda(w)))$ as $w \to +\infty$, and hence $I_2(w) = \mathcal{O}(\psi(\lambda(w)))$ as $w \to +\infty$.

We estimate $I_1(w)$.

$$\begin{aligned}I_1(w) &= \int_H L_w(s - h_w(t))\psi(|f(h_w(t)) - f(s)|)\,d\mu_H(t) \\ &\le \int_H L_w(s - h_w(t))\psi(R\,\Omega_w(s_w, t)\,\lambda(w))\,d\mu_H(t).\end{aligned}$$

Putting $B_w(s) = \int_H L_w(s - h_w(t))\,d\mu_H(t)$, by Lemma 8.1 and by assumption $(L_w.2)$, there exist $\overline{w} \in W$ and $r > 0$ such that $r < B_w(s) \le N$, for $w \in W$, $w > \overline{w}$ and for every $s \in G$. Hence by the concavity of $\psi \in \widetilde{\Psi}$, we have

$$\begin{aligned} I_1(w) &\le \frac{1}{B_w(s)} \int_H L_w(s - h_w(t)) B_w(s) \psi(R\,\Omega_w(s_w, t)\,\lambda(w))\, d\mu_H(t) \\ &\le N\psi\left(\frac{1}{B_w(s)} \int_H L_w(s - h_w(t)) R\,\Omega_w(s_w, t)\,\lambda(w)\, d\mu_H(t)\right) \\ &\le N\psi\left(\frac{R\lambda(w)}{r} \int_H L_w(s - h_w(t))\,\Omega_w(s_w, t)\, d\mu_H(t)\right) \\ &\le N\psi\left(\frac{R\lambda(w)}{r} m_\omega\right). \end{aligned}$$

Taking $r > 0$ so small that $Rm_\omega r^{-1} > 1$, again by the concavity of $\psi \in \widetilde{\Psi}$, there holds $I_1(w) = \mathcal{O}(\psi(\lambda(w)))$ as $w \to +\infty$, and therefore the assertion follows. □

8.5 Applications regarding rate of convergence

In this section we discuss some examples of operators (8.12) to which Theorem 8.6 can be applied.

(I) We consider, as in (I) of Section 8.3, the case of nonlinear convolution integral operators, i.e., when $G = H = (\mathbb{R}^N, +)$, $\mu_H = dt$ being the Lebesgue measure, and we put $h_w(t) = t$ for every $w \in W$; then we obtain

$$(T_w f)(s) = \int_{\mathbb{R}^N} K_w(s - t, f(t))\, dt, \quad s \in \mathbb{R}^N.$$

In this case we may take $\omega(x, y) = |x - y|^\alpha = w^\alpha |x - y|^\alpha w^{-\alpha} = (w|x - y|)^\alpha w^{-\alpha}$, $0 < \alpha \le 1$. So $\Omega_w(x, y) = \omega(w(x - y)) = (w|x - y|)^\alpha$ and $\lambda(w) = w^{-\alpha}$. Moreover

$$|f(s) - f(h_w(t))| = |f(s) - f(t)| \le R(w|s - t|)^\alpha w^{-\alpha},$$

and therefore $s_w = s \in \mathbb{R}^N$, for every $w \in W$. Furthermore,

$$r^w(s) = \sup_{u \ne 0} \left| \frac{1}{u} \int_{\mathbb{R}^N} K_w(s - t, u)\, dt - 1 \right| = \sup_{u \ne 0} \left| \frac{1}{u} \int_{\mathbb{R}^N} K_w(z, u)\, dz - 1 \right|$$

which is independent of s, and

$$\begin{aligned} m_\omega &= \sup_{w \in W} \sup_{s \in \mathbb{R}^N} \int_{\mathbb{R}^N} L_w(s - t)\omega(w(s - t))\, dt \\ &= \sup_{w \in W} \sup_{s \in \mathbb{R}^N} \int_{\mathbb{R}^N} L_w(s - t) w^\alpha |s - t|^\alpha\, dt \\ &= \sup_{w \in W} w^\alpha \int_{\mathbb{R}^N} L_w(z) |z|^\alpha\, dz \end{aligned}$$

represents the αth-moment of the kernel $\{L_w\}_{w\in W}$.

Hence Theorem 8.6 applied to the above family of nonlinear convolution integral operators can be formulated as follows.

Corollary 8.3. *Let* $f : \mathbb{R}^N \to \mathbb{R}$ *with* $f \in C(\mathbb{R}^N)$*, and let* $\{K_w\}_{w\in W} \subset \mathcal{K}_w$*,* $\{L_w\}_{w\in W} \subset \mathcal{L}_w$*, and* $\psi \in \widetilde{\Psi}$*. If* $f \in \mathrm{Lip}_\omega$*,* $m_\omega = \sup_{w\in W} w^\alpha \int_{\mathbb{R}^N} L_w(z)|z|^\alpha\, dz < +\infty$*, and*

$$r^w = \sup_{u\neq 0}\left|\frac{1}{u}\int_{\mathbb{R}^N} K_w(z,u)\,dz - 1\right| = \mathcal{O}(w^{-\alpha}) \quad \text{as } w \to +\infty,$$

then

$$\left\|\int_{\mathbb{R}^N} K_w(\cdot - t, f(t))\,dt - f(\cdot)\right\|_\infty = \mathcal{O}(\psi(w^{-\alpha})) \quad \text{as } w \to +\infty.$$

(II) We now take into consideration the case of discrete operators.

Here we put $G = (\mathbb{R}, +)$ and $H = (\mathbb{Z}, +)$ with the counting measure μ_H. We will use the same notations as in (III) of Section 8.3. For $\{K_w\} \subset \mathcal{K}_w$, the operators (8.12) now take on the form

$$(T_w f)(s) = \sum_{k=-\infty}^{+\infty} K_w(s - h_w(k), f(h_w(k))), \qquad s \in \mathbb{R},\ w \in W.$$

So we may take the function $\omega(x, y) = |x - y|^\alpha = \Omega_w(x, y)$, for every $w \in W$ and $0 < \alpha \le 1$; since $h_w(t) = h_w(k)$, we have

$$|f(s) - f(h_w(k)| \le R|s_w - k|^\alpha \lambda(w)$$

for some $s_w \in \mathbb{R}$ and $\lambda : \mathbb{R}^+ \to \mathbb{R}^+$ such that $\lim_{w\to+\infty} \lambda(w) = 0$. Moreover,

$$r^w(s) = \sup_{n\in\mathbb{N}} r_n^w(s) = \sup_{u\neq 0}\left|\frac{1}{u}\sum_{k=-\infty}^{+\infty} K_w(s - h_w(k), u) - 1\right|,$$

and

$$m_\omega = \sup_{w\in W}\sup_{s\in\mathbb{R}} \sum_{k=-\infty}^{+\infty} L_w(s - h_w(k))|s_w - k|^\alpha.$$

Now, let us consider the particular case of *nonlinear generalized sampling series*, i.e., when $W = \mathbb{R}^+$, $h_w(k) = \frac{k}{w}$, $w > 0$, $k \in \mathbb{Z}$, and $K_w(z, \cdot) = K(wz, \cdot)$, $w > 0$, $z \in \mathbb{R}$ and $K : \mathbb{R} \times \mathbb{R} \to \mathbb{R}$, were K satisfies a Lipschitz condition as in Example III of Section 8.3 and let $L_w(z) = L(wz)$, $w > 0$, $z \in \mathbb{R}$ where $L : \mathbb{R} \to \mathbb{R}_0^+$ is the measurable function in the Lipschitz condition of K. In this case

$$(T_w f)(s) = \sum_{k=-\infty}^{+\infty} K\left(ws - k, f\left(\frac{k}{w}\right)\right),$$

and

$$\omega(s, h_w(k)) = \left|s - \frac{k}{w}\right|^\alpha = \left|\frac{ws - k}{w}\right|^\alpha = |ws - k|^\alpha w^{-\alpha},$$

and hence $s_w = sw$, $\Omega_w(s_w, k) = |ws - k|^\alpha$ and $\lambda(w) = w^{-\alpha}$.

Therefore we obtain that the class Lip_ω consists of all functions $f : G \to \mathbb{R}$ such that there exists a constant $R > 0$ in such a way that, for every $x, y \in \mathbb{R}$,

$$|f(x) - f(y)| \le R|x - y|^\alpha,$$

and hence $|f(s) - f(h_w(k))| \le R|ws - k|^\alpha w^{-\alpha}$.

Moreover,

$$\begin{aligned} r^w(s) = \sup_{n\in\mathbb{N}} r_n^w(s) &= \sup_{u\neq 0}\left|\frac{1}{u}\sum_{k=-\infty}^{+\infty} K\left(w\left(s - \frac{k}{w}\right), u\right) - 1\right| \\ &= \sup_{u\neq 0}\left|\frac{1}{u}\sum_{k=-\infty}^{+\infty} K(ws - k, u) - 1\right|, \end{aligned}$$

and

$$\begin{aligned} m_\omega &= \sup_{w>0}\sup_{s\in\mathbb{R}}\sum_{k=-\infty}^{+\infty} L\left(w\left(s - \frac{k}{w}\right)\right)|ws - k|^\alpha \\ &= \sup_{w>0}\sup_{s\in\mathbb{R}}\sum_{k=-\infty}^{+\infty} L(ws - k)|ws - k|^\alpha. \end{aligned}$$

Note that $m_\omega < +\infty$ if $\sup_{s\in\mathbb{R}}\sum_{k=-\infty}^{+\infty} L(s - k)|s - k|^\alpha < +\infty$, and the latter is a classical assumption in the theory of generalized sampling series. Hence it is possible to state the following corollary.

Corollary 8.4. *Let $f : \mathbb{R} \to \mathbb{R}$, $f \in C(\mathbb{R})$, and let K be an (L, ψ)-Lipschitz kernel function, with $\psi \in \widetilde{\Psi}$ and where L satisfies assumption* b) *of* (III) *of Section* 8.3. *If $f \in \mathrm{Lip}_\omega$, $m_\omega < +\infty$, and*

$$r^w(s) = \sup_{u\neq 0}\left|\frac{1}{u}\sum_{k=-\infty}^{+\infty} K(ws - k, u) - 1\right| = \mathcal{O}(w^{-\alpha}), \quad \text{as } w \to +\infty,$$

uniformly with respect to $s \in \mathbb{R}$, with $0 < \alpha \le 1$, then

$$\left\|\sum_{k=-\infty}^{+\infty} K(w\cdot - k, f(\tfrac{k}{w})) - f(\cdot)\right\|_\infty = \mathcal{O}(\psi(w^{-\alpha})), \quad \text{as } w \to +\infty.$$

Remark 8.6. (a) We remark that if K is strongly-Lipschitz, then $\psi(u) = u$, for every $u \in \mathbb{R}_0^+$, and we obtain

$$\Big\| \sum_{k=-\infty}^{+\infty} K(w \cdot -k, f(\frac{k}{w})) - f(\cdot)\Big\|_\infty = \mathcal{O}(w^{-\alpha}), \qquad \text{as } w \to +\infty.$$

This result can be applied to the particular case of linear sampling operators, i.e., when $K(t,u) = \widetilde{K}(t)u$ (see [175]).

(b) According to Remark 8.5 (a), we may find a result on the order of approximation for the multivariate nonlinear sampling operator. In this case $\lambda : \mathbb{R}_+^N \to \mathbb{R}^+$ with $\lambda(w) = w^{-\alpha} = \big(\prod_{i=1}^N w_i\big)^{-\alpha}$, where $w = (w_1, w_2, \dots, w_N) \in W^N$.

8.6 Uniform regular methods of summability

In this section we will investigate the problem when the general family of operators (8.12) defines a $\mathbb{T}$-regular method of summability, as defined in Section 5.2, with respect to uniform convergence, i.e., the case in which both the convergence of the functions f_w and of the operators $T_w f_w$ is the uniform one. The results presented here are based on Theorem 8.5, and what is remarkable is that the classes $\mathcal{K}_w$ and $\mathcal{L}_w$ are exactly the same as considered there. Here the set of indices W will be taken as in the previous section. This results in a wholly unified theory.

Let $\{f_w\}_{w \in W}$ be a family of functions on G and consider the family of nonlinear operators $\mathbb{T} = (T_w)_{w\in W}$,

$$(T_w f_w)(s) = \int_H K_w(s - h_w(t), f_w(h_w(t)))\, d\mu_H(t),$$

defined for any $f_w \in \operatorname{Dom} \mathbb{T} = \bigcap_{w\in W} \operatorname{Dom} T_w$.

We say that $\{f_w\}_{w\in W} \subset C(G)$ is *uniformly* (T_w)*-summable* to $f \in C(G)$ if $T_w f_w \to f$ as $w \to +\infty$ uniformly with respect to $s \in G$.

We will say that $(T_w)_{w\in W}$ is a *uniform regular method of summability* on $C(G)$ if

$$\|f_w - f\|_\infty \to 0 \text{ as } w \to +\infty \text{ implies that } f_w \text{ is } T_w\text{-summable to } f,$$

i.e.,

$$\|T_w f_w - f\|_\infty \to 0, \qquad \text{as } w \to +\infty.$$

Now let $\mathcal{K}_w$ and $\mathcal{L}_w$ be families of kernels as defined in Section 8.2. We formulate the following

Theorem 8.7. *Let $f : G \to \mathbb{R}$, $f \in C(G)$, and suppose $\{K_w\}_{w\in W} \subset \mathcal{K}_w$ and $\{L_w\}_{w\in W} \subset \mathcal{L}_w$. Then $(T_w)_{w\in W}$ is a uniform regular method of summability on $C(G)$.*

Proof. Let $\{f_w\}_{w\in W}$ be a family of functions on $C(G)$ such that $\|f_w - f\|_\infty \to 0$ as $w \to +\infty$. We may write

$$\begin{aligned}|(T_w f_w)(s) - f(s)| &\leq |(T_w f_w)(s) - (T_w f)(s)| + |(T_w f)(s) - f(s)| \\ &= I_1^w(s) + I_2^w(s).\end{aligned}$$

We evaluate I_1^w. By the (L_w, ψ)-Lipschitz condition we have

$$\begin{aligned}I_1^w(s) &= \bigg| \int_H K_w(s - h_w(t), f_w(h_w(t)))\, d\mu_H(t) \\ &\qquad - \int_H K_w(s - h_w(t), f(h_w(t)))\, d\mu_H(t) \bigg| \\ &\leq \int_H L_w(s - h_w(t))\psi\left(|f_w(h_w(t)) - f(h_w(t))|\right) d\mu_H(t) \\ &\leq \int_H L_w(s - h_w(t))\psi(\|f_w - f\|_\infty)\, d\mu_H(t) \\ &\leq M\psi(\|f_w - f\|_\infty),\end{aligned}$$

and since $\psi \in \Psi$ and $\|f_w - f\|_\infty \to 0$ as $w \to +\infty$, we deduce that $I_1^w(s) \to 0$ as $w \to +\infty$ uniformly with respect to $s \in G$.

Now, $I_2^w(s) = |T_w f(s) - f(s)| \leq \|T_w f - f\|_\infty$, and so, by Theorem 8.5, $I_2^w(s) \to 0$ as $w \to +\infty$, uniformly with respect to $s \in G$.

Therefore the assertion follows, taking the supremum of $s \in G$. □

8.7 Bibliographical notes

It is not completely clear who first established the sampling theorem and in this respect there are different opinions. Indeed I. Kluvanek (1965, [127]) says: "The origin of this theorem can hardly be traced".

One of the historical roots of the sampling theorem is in interpolation theory with equidistant nodes; and considering the case of not necessarily band-limited functions, the first person who considered the sampling theorem in this respect was the Belgian mathematician Charles-Jean Baron de la Vallée Poussin in 1908 in [88]. His work dealt with the case of duration-limited functions, a class of functions which cannot be simultaneously band-limited and that represents signals which occur in practice.

De la Vallée Poussin considered the finite interpolation formula

$$F_m(x) = \sum_{k=-\infty}^{+\infty} f(a_k) \frac{\sin m(x - a_k)}{m(x - a_k)}$$

where f is a bounded function on a compact interval $[a, b]$ and $a_k := k\pi/m$, $k \in \mathbb{Z}$, $m = n$ or $m = n + 1/2$, $n \in \mathbb{N}$. Here it is assumed that $f : \mathbb{R} \to \mathbb{R}$ is zero outside

$[a, b]$. For such functions F_m, de la Vallée Poussin established sufficient conditions for pointwise and uniform convergence. But such a study is connected with the sampling theorem since F_m interpolates f at the nodes a_k.

De la Vallée Poussin's work was followed by M. Theis in 1919 (see [189]). Now in de la Vallée Poussin's approach continuity of a function f at a point x_0 alone does not suffice for the convergence of F_m to f, and further assumptions such as the existence of $f'(x_0)$ or the requirement that f is of bounded variation in a neighbourhood of x_0, are needed. However, in order to obtain convergence for any continuous functions f, Theis replaced the kernel $\frac{\sin x}{x}$ by $\left\{\frac{\sin x}{x}\right\}^2$, which is the counterpart of Fejér's method of summation of Fourier series, and she established a convergence result solely under continuity assumptions. Later on, in 1927, J. M. Whittaker in [207] generalized the convergence theorem for F_m of de la Vallée Poussin for functions which are not necessarily duration-limited. Studies following the same direction as those of de la Vallée Poussin include the work of J. L. Brown in 1967 (see [50]) with associates aliasing error estimates, and the work of P. L. Butzer's school at Aachen since 1976. As mentioned in [78], many electrical and communication engineers dealt with the de la Vallée Poussin interpolation formula even though they may not have been aware of the fact that he was one of its major initiators.

Underlying this kind of work, there was also interest in regarding the series F_m not in terms of its behavior for $m \to +\infty$, but from the point of view of an interpolation formula. In this respect the paper [206] of E. T. Whittaker in 1915 contains the interpolation problem of finding a function passing through the points $\left(\frac{k}{W}, f\left(\frac{k}{W}\right)\right)$, $k \in \mathbb{Z}$, $W > 0$. Among all analytic functions which are solutions of the above problem, he choose the function

$$C(x) := \sum_{k=-\infty}^{+\infty} f\left(\frac{k}{W}\right) \frac{\sin \pi(Wx - k)}{\pi(Wx - k)}.$$

Note that $C(x)$ is just $F_m(x)$ for $m = \pi W$, and $C(x)$ is an entire function, band-limited to $[-\pi W, \pi W]$. The above series, called by Whittaker cardinal series, can be obtained as a limiting case of Lagrange's interpolation formula as the number of nodes tends to infinity. A number of mathematicians have studied the relationship of the above series with the Newton–Gauss series, the Everett, Stirling and Bessel series; all these series solve the same interpolation problem and are related to the cardinal series.

The interpolation problem posed in the sampling theorem, as shown in Section 8.1, is the so-called Shannon sampling theorem for band-limited functions. Indeed C. E. Shannon established the following important engineering principle: *If a signal f has a bounded frequency content, then all the information contained in such a signal is in fact contained in the sample values at equidistantly spaced sample points and knowledge of the bound determines the minimum rate at which the signal needs to be sampled in order to reconstruct it exactly. This rate, which is called the "Nyquist sampling rate", is measured in W samples per second.*

Shannon's work goes on to discover that other sets of data can be used in order to reconstruct the band-limited signal f: namely, one can also consider the values of f and its first derivative at every other sample point, or the values of f and its first and second derivative at every third sample point and further, it is also possible to use sample points which are not uniformly spaced (non uniform or irregular sampling). An interesting example which illustrates a concrete sampling situation in which the use of the derivative is important, is given in [118]. There, as an example, is discussed an airplane pilot's instruments panel, where derivative information could be available to the pilot. Moreover, even if a version of this theorem was known by Borel in 1897, the sampling theorem was introduced into the engineering literature by Shannon, whose paper was apparently written in 1940 but not published until after World War II in 1949 ([179]); but it seems that its contents were in circulation in the United States by 1948 (see [179]).

A little later, news emerged from Russia that Kotel'nikov had published the sampling theorem in 1933 ([130]), and there it was known by his name. His results began to appear in the western literature in the late 1950s. Other contribution to sampling theory were given by Someya in [183], who continued the Japanese interest in cardinal series going back to Ogura in 1920 ([168]), and by Weston in [205]. Other contributors to the engineering literature were Nyquist, Bennet, Gabor and Raabe. But it is really impossible to quote here all the mathematicians who have been interested in the sampling theorem directly or not; indeed we invite the interested reader to read the survey papers by Jerri [122], Butzer [55], Butzer–Stens [78] and Higgins [118].

If now we denote by $(S_W f) := C(x)$, and we observe that $(S_{m/\pi} f)(x) = F_m(x)$, then it transpires that de la Vallée Poussin was the first who in 1908 considered the reconstruction of a function f by means of its sampling series, being f duration-limited, (i.e. f has compact support).

The interpolation formula of the sampling theorem shown in Section 8.1, has been considered by many authors, but in Ogura's paper ([168]) of 1920 there is a formulation of the sampling theorem similar to Shannon's' (Section 8.1). Even so, Ogura attributes it to E. T. Whittaker (1915), probably erroneously. Ogura seems to be the first to prove the sampling theorem rigorously, and J. M. Whittaker in 1927 ([207]) obtained results containing a weak version of the sampling theorem. But probably, neither Ogura, nor J. M. Whittaker, unlike V. A. Kotel'nikov and C. E. Shannon, realized the importance of their results in relation to the sampling theorem.

Another important step in the history of the sampling theorem appeared when in 1960's some mathematicians began to consider the interpolation formula of the classical sampling theorem when the band-limitation is weakened in some sense. Results in this respect are due to P. Weiss (1965, [204]) and J. L. Brown (1967, [50]), and gave the following estimates.

If $f \in L^2(\mathbb{R}) \cap C(\mathbb{R})$ is such that $\hat{f} \in L^1(\mathbb{R})$, then there holds:

$$|f(x) - (S_W f)(x)| \leq \sqrt{\frac{2}{\pi}} \int_{|v| \geq \pi W} |\hat{f}(v)|\, dv, \qquad W > 0.$$

The above inequality is interesting since the right-hand side tends to zero for $W \to +\infty$ and this shows that f can be uniformly approximated by the sampling series $S_W f$. This also shows that f is not exactly represented by the interpolation series as in the sampling theorem, but it can only be approximated when $W \to +\infty$; and this goes in the same direction of the work of de la Vallée Poussin. Moreover, the above upper bound for the error occurring when f is approximated by $S_W f$, is the best possible in the sense that the constant $\sqrt{\frac{2}{\pi}}$ cannot be improved. Further, it is important to note that in the particular case of a band-limited f, the right-hand side vanishes and we obtain the exact representation of f by means of $S_W f$, which is the Shannon sampling theorem.

This is just in the direction of the work of P. L. Butzer and his school at Aachen. Earlier they considered time (duration)-limited functions and later this condition was weakened assuming conditions on the behavior of f at infinity. Later they began to consider the generalized sampling series of f, obtaining the results quoted in Section 8.1 (see [79]). Concerning the behavior of the function f in correspondence of jump-discontinuities, they gave a partial solution to a conjecture of R. Bojanic. They proved that a series having the interpolation property cannot converge at points of discontinuity. But if the assumption that the kernel φ in $S_W f$ is continuous, is dropped, then in this case they showed that there exist kernels φ such that $(S_W f)(x)$ interpolates f at the nodes n/W, $n \in \mathbb{Z}$ and converges as $W \to +\infty$.

Subsequently C. Bardaro and G. Vinti in [36] considered the generalized sampling series in the sense of Butzer and in [39], for the first time, the case of a generalized sampling series in its nonlinear form was studied and uniform approximation results and error estimates proved. Here the authors used a general approach in a locally compact topological group, as did Kluvanek in [127] and later Beaty and Dodson in [120], [44] for the classical Shannon sampling theorem. More precisely, the authors considered a general form of linear or nonlinear integral operators from which it is possible to deduce the generalized sampling case, as also other classical cases, as convolution or Mellin operators. Moreover in [40], [41] the authors also investigated the possibility that the convergence process for the family of nonlinear generalized sampling operators or more generally for the family of nonlinear integral operators considered in Section 8.2, defines a regular method of summability in the sense mentioned there.

To conclude, even if the origin of sampling theorem seems to be a very complicated problem, Butzer and Stens ([79]) say that de la Vallée Poussin can be considered as the father of sampling theory in the case of time-limited functions. Moreover M. Theis was the first to consider generalized sampling series in the sense that the kernel $\sin x/x$ is replaced by the particular Fejér kernel $(\sin \pi x/\pi x)^2$.

Chapter 9

Modular approximation by sampling type operators

As seen in Section 8.1, the classical sampling theorem holds for continuous and band-limited signals which have finite energy, i.e., for functions $f \in L^2(\mathbb{R}) \cap C^0(\mathbb{R})$ (being $C^0(\mathbb{R})$ the space of continuous functions on $\mathbb{R}$), while in Theorem 8.1, in a different frame, it is required for the generalized sampling operator that $f \in C(\mathbb{R})$. In Section 8.2 the results of Section 8.1 have been extended to the nonlinear case still in the frame of $C(\mathbb{R})$. Now coming back for a moment to the classical signal analysis, it would be of some interest to give an approximation result, in the nonlinear frame of Section 8.2, for functions belonging only to $L^2(G)$, i.e. for signals having finite energy, but which are not necessarily either continuous or band-limited. But still more than this, it would be of some interest to give such approximation results for functions belonging to an L^p-space, for $p \geq 1$ or to a more general functional space. Indeed, this fact may have the following interpretation in signal theory: the power of a signal is defined as

$$P_f = \lim_{T \to +\infty} \frac{1}{2T} \int_{-T}^{T} |f(t)|^2 \, dt,$$

and there exist signals which have finite power, but infinite energy, which means that one can deal with signals which do not belong to L^2. As examples of these kind of signals there are the periodic power signals, the aperiodic ones and the random processes.

9.1 Modular convergence for a class of nonlinear integral operators

In this section, we will develop a theory of convergence for operators (8.12) in modular spaces. As seen in Section 1.2, the theory of modular spaces contains, in particular, that one of the Musielak–Orlicz and the Orlicz spaces, which are generalizations of the classical L^p-spaces. In this manner, we have a theory that, from the point of view of applications, gives a unifying approach to signals having finite energy, but also to other signals like, for example, the power signals.

The concept of convergence used is the modular convergence introduced in Section 1.2.

We will use the notions of Section 8.2, and we will deal with the family of nonlinear integral operators (8.12). Let μ_G be the Haar measure on the class of Borel sets $\mathcal{B}(G)$, associated to the group G.

By $L^0(G)$ and $L^0(H)$ we will denote the vector spaces of all Borel measurable real-valued functions defined on G and H, respectively, and by $L^0_\rho(G)$ and $L^0_\rho(H)$ the corresponding modular spaces. Moreover, by $C_c(G)$ we denote the subspace of all continuous functions with compact support on G. For $G = (\mathbb{R}^n, +)$ we denote by $C_c^{(r)}(\mathbb{R}^n)$ the space of all real-valued functions with compact support and with continuous derivative of order r, for $1 \leq r \leq +\infty$.

The general setting of modular spaces has required the use of some notions on the modulars taken into consideration, which we have introduced in Section 2.1. A new notion is needed in order to formulate Theorems 9.2 and 9.4, which is the following.

We say that a modular ρ is *strongly finite* if the characteristic function χ_A of a set A of finite measure belongs to the modular space E_ρ.

Of course, such notion, like finiteness, absolute finiteness, absolute continuity and monotonicity, is obviously satisfied when one deals with modulars which have an integral representation like, for example, the modulars generating the Orlicz spaces. In case of modulars generating a Musielak–Orlicz space, we recall that finiteness and absolute finiteness or strong finiteness are equivalent to the requirement that the φ-function generating the modular is locally integrable for small u and is locally integrable, respectively (see Example 2.1 (b) for the above notions). Moreover it is easy to show that, if the underlying space G is of finite measure, (for example if G is a compact group), then absolute continuity together with strong finiteness implies absolute finiteness (see [16]).

Let W be an unbounded subset of $\mathbb{R}^+$, as in Section 8.2. Now let $(G, \mathcal{B}(G), \mu_G)$ and $(H, \mathcal{B}(H), \mu_H)$ be two locally compact and σ-compact topological groups and let $\{\mu_w\}_{w\in W}$ be a family of functions $\mu_w : G \times \mathcal{B}(H) \to \mathbb{R}_0^+$, with $w \in W$ such that $\mu_w(\cdot, A)$ is measurable for every $A \in \mathcal{B}(H)$ and $\mu_w(s, \cdot)$ is a measure on $\mathcal{B}(H)$.

Now we introduce a notion of "regularity" on the family $\{\mu_w\}_{w\in W}$.

We will say that $\{\mu_w\}_{w\in W}$ is *regular* if

a) putting $\mu_w^s = \mu_w(s, \cdot)$, for $w \in W, s \in G$ we have $0 \leq \mu_w^s \ll \mu_H$ for every $s \in G$, $w \in W$, where "$\ll$" means that μ_w^s is absolutely continuous with respect to μ_H,

b) $\sup_{w\in W} \|\mu_w^{(\cdot)}(H)\|_\infty < +\infty$ and there exist a measurable set $F \subset G$ with $\mu_G(F) = 0$ and two positive real constants r and $\overline{w} \in W$ such that for every $w \in W$, $w > \overline{w}$ and for every $s \in G \setminus F$, one has $0 < r \leq \mu_w^{(s)}(H)$,

c) setting $\xi_w(s, t) = \dfrac{d\mu_w^s}{d\mu_H}$, we have that ξ_w is a globally measurable function and $\|\xi_w(\cdot, t)\|_1 \leq \eta_w$ for every $t \in H$ and for $\eta_w \in \mathbb{R}^+$.

Let now ρ_G, ρ_H be two modular functionals on $L^0(G)$ and $L^0(H)$, respectively. Due to the general setting of modular spaces, we introduce a condition of *compatibility between the family* $\{\mu_w\}_{w\in W}$ *and the modulars* ρ_G *and* ρ_H.

We will say that a regular net $\{\mu_w\}_{w\in W}$ is *compatible with the couple* (ρ_G, ρ_H) if there are two constants $D, M > 0$ and a net $\{b_w\}$ of positive numbers with $b_w \to 0$ as $w \to +\infty$ such that

$$\rho_G\left(\int_H g(t)\, d\mu_w^{(\cdot)}(t)\right) \leq M\eta_w \rho_H(Dg) + b_w$$

for any $g \geq 0$, $g \in L^0(H)$ and for sufficiently large $w \in W$, where η_w is the net introduced in assumption c) of the regularity of $\{\mu_w\}_{w\in W}$.

Now we will show that if ρ_G, ρ_H are convex modulars generating Orlicz spaces, then every regular net $\{\mu_w\}_{w\in W}$ is compatible with the couple (ρ_G, ρ_H).

Proposition 9.1. *Let* $\varphi : \mathbb{R}_0^+ \to \mathbb{R}_0^+$ *be a convex* φ*-function as defined in Example* 1.5 (b), *and let*

$$I_\varphi^G(f) := \rho_G(f) = \int_G \varphi(|f(t)|)\, d\mu_G(t),$$

and

$$I_\varphi^H(g) := \rho_H(g) = \int_H \varphi(|g(t)|)\, d\mu_H(t)$$

be the modulars on $L^0(G)$ *and* $L^0(H)$ *generating the Orlicz spaces. Then every regular net* $\{\mu_w\}_{w\in W}$ *is compatible with the couple* (ρ_G, ρ_H).

Proof. Let us consider a regular family $\{\mu_w\}_{w\in W}$ according to the previous definitions. For every $s \geq 0$, putting

$$\beta_w(s) = \mu_w(s, H) = \mu_w^{(s)}(H)$$

and $\sup_{w\in W} \| \mu_w^{(\cdot)}(H) \|_\infty \equiv D$ we have $\beta_w(s) \leq D$ for every $s \in G$, $w \in W$ and

$$\rho_G\left[\int_H g(t)\, d\mu_w^{(\cdot)}(t)\right] = \rho_G\left[\int_H g(t)\beta_w(s)\, d\widetilde{\mu}_w^{(\cdot)}(t)\right]$$

where $\widetilde{\mu}_w^{(s)}(E) := \frac{\mu_w^{(s)}(E)}{\beta_w(s)}$, $E \in \mathcal{B}(H)$. Hence, by using Jensen's inequality, we obtain

$$\begin{aligned}
\rho_G\left[\int_H g(t)\, d\mu_w^{(\cdot)}(t)\right] &= \int_G \varphi\left[\int_H g(t)\, d\mu_w^{(\cdot)}(t)\right] d\mu_G(s) \\
&\leq \int_G \left\{\int_H \varphi[\beta_w(s)g(t)]\, d\widetilde{\mu}_w^{(s)}(t)\right\} d\mu_G(s) \\
&= \int_G \left\{\int_H \frac{1}{\beta_w(s)} \varphi[\beta_w(s)g(t)]\, d\mu_w^{(s)}(t)\right\} d\mu_G(s).
\end{aligned}$$

Since $\mu_w(s,\cdot) \ll \mu_H$, there exists a nonnegative measurable function $\xi_w(s,t)$, such that $d\mu_w(s,\cdot) = \xi_w(s,t)d\mu_H(\cdot)$. Thus, by the definition of regularity of $\{\mu_w\}$, we deduce

$$\begin{aligned}\rho_G[\int_H g(t)\,d\mu_w^{(\cdot)}(t)] &\leq \frac{1}{r}\int_G \left\{\int_H \varphi[Dg(t)]\xi_w(s,t)\,d\mu_H(t)\right\} d\mu_G(s)\\ &= \frac{1}{r}\int_H \varphi[Dg(t)]\left\{\int_G \xi_w(s,t)\,d\mu_G(s)\right\} d\mu_H(t)\\ &\leq \frac{\eta_w}{r}\rho_H(Dg)\end{aligned}$$

for sufficiently large $w \in W$. So the assertion follows with $b_w \equiv 0$ and $M = 1/r$. □

Concerning the assumptions on the families $\{K_w\}_{w\in W}$ and $\{L_w\}_{w\in W}$ we will assume that $\{K_w\}_{w\in W} \subset \mathcal{K}_w$ where in the assumption (K_w.3) of Section 8.2 we will use the weaker requirement that $\lim_{w\to+\infty} r_n^w(s) = 0$ a.e. on G, and the class $\mathcal{L}_w$ is given by the family of measurable functions $L_w : G \to \mathbb{R}_0^+$ such that $L_w(s-h_w(\cdot)) \in L^1_{\mu_H}(H)$ for $s \in G$, satisfying the following assumptions:

(L_w.1) $L_w : G \to \mathbb{R}_0^+$ are functions with compact support on G such that for every $U \in \mathcal{U}$ there is a $\overline{w} \in W$ such that for every $w \in W$, $w > \overline{w}$ the supports of L_w are contained in U,

(L_w.2) there is a constant $N > 0$ such that

$$\int_H L_w(s-h_w(t))\,d\mu_H(t) \leq N,$$

for every $s \in G$ and $w \in W$,

(L_w.3) there are measurable set $F \subset G$ with $\mu_G(F) = 0$ and two positive real constants r and $\overline{w} \in W$ such that for every $w \in W$ $w > \overline{w}$ and for every $s \in G \setminus F$, we have

$$\int_H L_w(s-h_w(t))\,d\mu_H(t) > r.$$

We remark that the new assumption (L_w.1) implies that one of Section 8.2. Indeed, we may establish the following

Proposition 9.2. *If (L_w.1) is satisfied, then for every $U \in \mathcal{U}$ and $s \in G$*

$$\lim_{w\to+\infty} \int_{H\setminus h_w^{-1}(s+U)} L_w(s-h_w(t))\,d\mu_H(t) = 0.$$

Proof. Let $U \in \mathcal{U}$ be fixed. From $(L_w.1)$ there exists $\overline{w} \in W$ such that $\operatorname{supp} L_w \subset U$ for $w \in W, w \geq \overline{w}$. If $t \in H \setminus h_w^{-1}(s+U)$, then for every $z \in U$ we have that $t \neq h_w^{-1}(s+z)$, for every $s \in G$; since h_w is a homeomorphism, for every $w \in W$, we may conclude that $h_w(t) \neq s+z$, and hence $s - h_w(t) \neq -z$, which implies that $s - h_w(t) \notin U$. Now, from the assumption $(L_w.1)$ we obtain for $w \geq \overline{w}$ that

$$\int_{H \setminus h_w^{-1}(s+U)} L_w(s - h_w(t))\, d\mu_H(t) = 0, \qquad \text{for every } s \in G,$$

i.e. the assertion follows. □

Let moreover $\mathcal{L}_w^* \subset \mathcal{L}_w$ be a subclass of $\mathcal{L}_w$ whose elements $\{L_w\}_{w\in W}$ satisfy the further condition that $L_w \in L^1_{\mu_G}(G)$ for every $w \in W$, and $\gamma_w := \int_G L_w(z)\, d\mu_G(z)$ for $w \in W$, is a bounded net.

It is clear that if in particular we take $\widetilde{\mu}_w^s(A) := \int_A L_w(s - h_w(t))\, d\mu_H(t)$, $A \in \mathcal{B}(H)$, $s \in G$, $w \in W$, then the family $\{\widetilde{\mu}_w\}_{w\in W}$ is regular.

Indeed, the condition a) of the regularity is obviously satisfied and b) is a consequence of $(L_w.2)$ and $(L_w.3)$. Finally, assumption c) is satisfied with $\eta_w = \gamma_w$, $w \in W$.

Now, we show some sufficient conditions under which the compatibility of a regular net with the couple (ρ_G, ρ_H) holds, if ρ_G, ρ_H are modulars generating Musielak–Orlicz spaces.

We take G and H such that $H \subset G$ (i.e., H is a subgroup of G) and μ_H will denote the Haar measure of the subgroup H. Let now $\varphi : G \times \mathbb{R}_0^+ \to \mathbb{R}_0^+$ be a measurable function with respect to $s \in G$, for every $u \in \mathbb{R}_0^+$, and such that $\varphi(s, \cdot)$ is a convex φ-function, as in Example 1.5 (c), which satisfies inequality (1.9) of Section 1.4 in the following equivalent form.

There are a constant $C \geq 1$ and a globally measurable function $F : G \times G \to \mathbb{R}_0^+$ such that for every $t, s \in G$ and $u \geq 0$ there holds

$$\varphi(s, u) \leq \varphi(t, Cu) + F(t, s). \tag{9.1}$$

Let $\{\widetilde{\mu}_w\}_w$ be the family defined by means of the net of kernels $\{L_w\}_{w\in W} \subset \mathcal{L}_w$. We set, according to Example 1.10 (b),

$$I_\varphi^G(f) := \rho_G(f) = \int_G \varphi(s, |f(s)|)\, d\mu_G(s),$$
$$I_\varphi^H(g) := \rho_H(g) = \int_H \varphi(t, |g(t)|)\, d\mu_H(t)$$

for $f \in L^0_{\rho_G}(G)$ and $g \in L^0_{\rho_H}(H)$. Note that inequality (9.1) implies the subboundedness of the modulars ρ_G and ρ_H (see Example 1.10 (b)). Finally we put

$$\delta_w = \int_G \int_H F(t, s) L_w(s - h_w(t))\, d\mu_H(t) d\mu_G(s).$$

Then we may prove the following

Proposition 9.3. *If $\delta_w \to 0$ as $w \to +\infty$, the family $\{\widetilde{\mu}_w\}_w$ is compatible with the couple (ρ_G, ρ_H).*

Proof. Putting $\widetilde{\mu}_w(s, E) = \widetilde{\mu}_w^s(E) = \int_E L_w(s - h_w(t))\, d\mu_H(t)$, $E \in \mathcal{B}(H)$, $s \in G$, we have

$$\rho_G\Big[\int_H g(t)\, d\widetilde{\mu}_w^{(\cdot)}(t)\Big] = \int_G \varphi\Big[s, \int_H g(t)\, d\widetilde{\mu}_w^s(t)\Big]\, d\mu_G(s).$$

By the Jensen inequality and $(L_w.2)$

$$\rho_G\Big[\int_H g(t)\, d\widetilde{\mu}_w^{(\cdot)}(t)\Big] \le \frac{1}{r}\int_G \Big\{\int_H \varphi(s, Ng(t)) L_w(s - h_w(t))\, d\mu_H\Big\}\, d\mu_G(s).$$

Now applying inequality (9.1) to the function φ, we obtain

$$\begin{aligned}\rho_G\Big[\int_H g(t)\, d\widetilde{\mu}_w^{(\cdot)}(t)\Big] \le \frac{1}{r}\int_G \Big\{\int_H \varphi[t, NCg(t)] L_w(s - h_w(t))\, d\mu_H(t)\Big\}\, d\mu_G(s)\\ + \frac{1}{r}\int_G \Big\{\int_H F(t, s) L_w(s - h_w(t))\, d\mu_H(t)\Big\}\, d\mu_G(s).\end{aligned}$$

By the Fubini–Tonelli theorem and since $\{L_w\}_{w\in W} \subset \mathcal{L}_w^*$, we obtain

$$\rho_G\Big(\int_H g(t)\, d\widetilde{\mu}_w^{(\cdot)}(t)\Big) \le \frac{1}{r}\gamma_w \rho_H(NCg) + \frac{\delta_w}{r},$$

and so the assertion follows, with $M = 1/r$, $\gamma_w = \eta_w$, $D = NC$ and $b_w = \delta_w / r$. □

Here are some examples in which the condition $\delta_w \to 0$ as $w \to +\infty$ is satisfied.

Example 9.1. (I) Let $G = H = (\mathbb{R}, +)$ and $h_w(t) = t$, with the Lebesgue measure. Let us suppose that the function φ satisfies (1.9) in the following strong sense.

There is a constant $C \ge 1$ and a globally measurable function $\Gamma : \mathbb{R} \times \mathbb{R} \to \mathbb{R}_0^+$ such that

$$\varphi(t - s, u) \le \varphi(t, Cu) + \Gamma(t, s)$$

for every $t, s \in G, u \in \mathbb{R}_0^+$ where Γ is such that $||\Gamma(\cdot, s)||_{L^1(\mathbb{R})} =: Q(s) \to 0$ as $s \to 0$.

In this case we have

$$\varphi(s, u) \le \varphi(t, Cu) + \Gamma(t, t - s),$$

and so we can put $F(t, s) = \Gamma(t, t - s)$.

Now we calculate δ_w:

$$\begin{aligned}\delta_w &= \int_{\mathbb{R}}\int_{\mathbb{R}} F(t,s)L_w(s-t)\,dt\,ds \\ &= \int_{\mathbb{R}}\int_{\mathbb{R}} \Gamma(t,t-s)L_w(s-t)\,dt\,ds \\ &= \int_{\mathbb{R}}\left\{\int_{\mathbb{R}} \Gamma(t,-(s-t))L_w(s-t)\,ds\right\}dt \\ &= \int_{\mathbb{R}}\left\{\int_{\mathbb{R}} \Gamma(t,-z)L_w(z)\,dz\right\}dt \\ &= \int_{\mathbb{R}} L_w(z)\left\{\int_{\mathbb{R}} \Gamma(t,-z)\,dt\right\}dz.\end{aligned}$$

Let $\varepsilon > 0$ be fixed. Then there is $\delta > 0$ such that $Q(-z) < \varepsilon$ for $|z| < \delta$. Moreover, there is $\widetilde{w} \in W$ such that for every $w \in W$, $w > \widetilde{w}$,

$$\delta_w = \int_{-\delta}^{\delta} L_w(z)Q(-z)\,dz < \varepsilon \int_{\mathbb{R}} L_w(z)\,dz,$$

and if $||L_w||_{L^1(\mathbb{R})}$ is a bounded net, we have $\delta_w \to 0$, as $w \to +\infty$.

(II) Let $G = (\mathbb{R},+)$, $H = (\mathbb{Z},+)$ with μ_G the Lebesgue measure on $\mathbb{R}$, μ_H the counting measure on $\mathbb{Z}$ and take $h_w(k) = \frac{k}{w}$. Moreover, we put, for $f \in L^0(\mathbb{R})$ and $g : \mathbb{Z} \to \mathbb{R}$,

$$I_\varphi^G(f) := \rho_G(f) = \int_{\mathbb{R}} \varphi(s,|f(s)|)\,ds,$$

and

$$I_\varphi^H(g) := \rho_H(g) = \sum_{k=-\infty}^{+\infty} \varphi(k,|g(k)|).$$

By inequality (9.1), we obtain

$$\varphi(s,u) \le \varphi(k,Cu) + F(k,s)$$

where $F(k,s) = \Gamma(k,k-s)$. In this case,

$$\delta_w = \int_{\mathbb{R}} \sum_{k=-\infty}^{+\infty} F(k,s)L_w\left(s - \frac{k}{w}\right)ds.$$

If we take $L_w(s) = L(ws)$, for $L \in L^1(\mathbb{R})$ and $w \in W$, we obtain

$$\begin{aligned}\delta_w &= \sum_{k=-\infty}^{+\infty} \int_{\mathbb{R}} F(k,s)L(ws-k)\,ds \\ &= \sum_{k=-\infty}^{+\infty} \int_{\mathbb{R}} \Gamma(k,k-s)L(ws-k)\,ds \\ &= \sum_{k=-\infty}^{+\infty} \frac{1}{w}\int_{\mathbb{R}} \Gamma\Big(k, \frac{kw-k-z}{w}\Big)L(z)\,dz \\ &= \frac{1}{w}\int_{\mathbb{R}} \sum_{k=-\infty}^{+\infty} \Gamma\Big(k, \frac{kw-k-z}{w}\Big)L(z)\,dz.\end{aligned}$$

Therefore it is sufficient to assume that

$$Q(s) := ||\Gamma(\cdot, s)||_{l^1(\mathbb{Z})}$$

is uniformly bounded with respect to $s \in \mathbb{R}$ in order to get the required result.

In order to state the main modular approximation results for the operators (8.12), we will use a density result which we state as follows.

Theorem 9.1. *Let ρ be a modular on $L^0(G)$, absolutely continuous, monotone and absolutely finite. Then $\overline{C_c(G)}^{\rho} = L^0_\rho(G)$ where "$\overline{}^{\rho}$" represents the modular closure.*

Proof. Let $a > 0$ be an absolute constant and consider a characteristic function χ_A with $A \in \mathcal{B}(G)$, $\mu_G(A) < +\infty$. Let $\varepsilon > 0$ be fixed. Then, since ρ is absolutely finite, there exists $\delta_{\varepsilon,a}$ such that we have $\rho(a\chi_D) < \varepsilon$ for $D \in \mathcal{B}(G)$ and $\mu_G(D) < \delta_{\varepsilon,a}$. Since G is σ-compact, the measure μ_G is regular. Then there are a compact set $C \subset A$ and an open set $V \supset A$ such that $\mu_G(V \setminus C) < \delta_{\varepsilon,a}$.

By the locally compact version of Urysohn's lemma, there is $f \in C_c(G)$ such that $\chi_A \le f \le \chi_V$, and so by monotonicity of ρ_G,

$$\rho_G(a(\chi_A - f)) \le \rho_G(a\chi_{V\setminus C}) < \varepsilon.$$

Let now $\xi : G \to \mathbb{R}$ be a simple function, that is $\xi(s) = \sum_{k=1}^N a_k\chi_{A_k}(s)$ with $\mu_G(A_k) < +\infty$ and $A_m \cap A_n = \emptyset$, $n \neq m$. Let ε_j be a sequence with $\varepsilon_j \downarrow 0$. For every j and $k = 1, 2, \dots, N$ there is a function $f_{j,k} \in C_c(G)$ such that $\rho_G[\Lambda(\chi_{A_k} - f_{j,k})] \le \frac{\varepsilon_j}{N}$, $k = 1, \dots, N$, where $\Lambda = \sum_{k=1}^N |a_k|$. Consider $f_j(s) = \sum_{k=1}^N a_k f_{j,k}(s)$. Then we have $f_j \in C_c(G)$ and

$$\rho_G(\xi - f_j) \le \rho_G\Big(\sum_{k=1}^N |a_k||\chi_{A_k} - f_{j,k}|\Big) \le \sum_{k=1}^N \rho_G[\Lambda|\chi_{A_k} - f_{j,k}|] < \varepsilon_j.$$

Thus for every simple function ξ there is a sequence of $f_j \in C_c(G)$ with $\rho_G(\xi - f_j) < \varepsilon_j$, that is there exists a sequence $\{f_j\} \subset C_c(G)$ with $\lim_{j\to+\infty} \rho_G(\xi - f_j) = 0$. Let $f \in L^0_\rho(G)$ be fixed.There exists a constant $\gamma > 0$ such that $\rho_G(3\gamma f) < +\infty$. If $f \geq 0$, there exists a monotone sequence of simple functions ξ_k such that $\xi_k \nearrow f$ a.e. Moreover $0 \leq f - \xi_k \leq f$ and so by the Lebesgue dominated convergence theorem for modular spaces formulated in Section 2.1 we have $\rho_G(\lambda(f - \xi_k)) \to 0$ with $\lambda \leq \gamma$. For general f we consider $f = f^+ - f^-$. In this case it is sufficient to apply the previous method separately to f^+ and f^-.

We can suppose $\lambda < 1$. From the above reasoning, for every ξ_k there is a sequence $\{f_{j,k}\}_j \subset C_c(G)$ with $\lim_{j\to+\infty} \rho_G(\xi_k - f_{j,k}) = 0$. Now we consider the function $\nu : \mathbb{N} \to \mathbb{N}$ defined as follows: for every $k \in \mathbb{N}$, $\nu(k)$ is the first integer such that $\nu(k) > k$ and $\rho_G(\xi_k - f_{\nu(k),k}) < \frac{1}{k}$. Finally we put $f_k = f_{\nu(k),k}$. Then $f_k \in C_c(G)$ and

$$\begin{aligned} \rho_G(\frac{\lambda}{2}(f - f_k)) &\leq \rho_G[\lambda(f - \xi_k)] + \rho_G[\lambda(\xi_k - f_k)] \\ &\leq \rho_G[\lambda(f - \xi_k)] + \rho_G[\xi_k - f_k] \\ &< \frac{1}{k} + \rho_G[\lambda(f - \xi_k)]. \end{aligned}$$

Thus $\lim_{k\to+\infty} \rho_G\left[\frac{\lambda}{2}(f - f_k)\right] = 0$ and the assertion follows. □

If $G = (\mathbb{R}^N, +)$ the space $C_c(G)$ can be replaced by $C_c^\infty(\mathbb{R}^N)$ by using a C^∞-version of Urysohn lemma.

Now we will assume that ρ_H and η_H are two modulars on $L^0(H)$ such that $\{\rho_H, \psi, \eta_H\}$ is a properly directed triple (see Section 1.4; here for the sake of simplicity we will suppose $\Omega = \Omega_0 = H$).

Finally, for $\{K_w\}_{w\in W} \in \mathcal{K}_w$, we define the operators

$$(T_w f)(s) = \int_H K_w(s - h_w(t), f(h_w(t)))\, d\mu_H(t)$$

for every $f \in \operatorname{Dom} \mathbb{T} = \bigcap_{w\in W} \operatorname{Dom} T_w$.

We will prove the following corollary, which will be used afterwards.

Corollary 9.1. *For every open set $A \subset G$ there is a $\overline{w} \in W$ such that for every $w \in W$, $w \geq \overline{w}$ there results $h_w^{-1}(A) \neq \emptyset$.*

Proof. If the assertion is false, there are an open set $A_0 \subset G$ and a sequence of positive numbers w_n, $n \in \mathbb{N}$, such that $h_{w_n}^{-1}(A_0) = \emptyset$. Let $f \in C_c(A_0)$ be not identically zero and let us set $C = \operatorname{supp} f$. Consider

$$C_n = \{t \in H : h_{w_n}(t) \in C\} = h_{w_n}^{-1}(C).$$

Then $C_n = \emptyset$ and so $T_{w_n} f \equiv 0$. By Theorem 8.5 this means that $f \equiv 0$, a contradiction. □

Now we state the following modular approximation theorem for functions belonging to $C_c(G)$.

Theorem 9.2. *Let* $\{K_w\}_{w\in W} \subset \mathcal{K}_w$ *and* $\{L_w\}_{w\in W} \subset \mathcal{L}_w$*; let* ρ *be a modular on* $L^0(G)$*, monotone, strongly finite and absolutely continuous. Then for every* $f \in C_c(G)$ *and* $\lambda > 0$*, there results*

$$\lim_{w\to+\infty} \rho[\lambda(T_w f - f)] = 0.$$

Proof. Let $f \in C_c(G)$ and let $C = \operatorname{supp} f$; we will suppose that $C \neq \emptyset$. Moreover we put

$$C_w = \{t \in H : h_w(t) \in C\} = h_w^{-1}(C).$$

By Corollary 9.1, C_w is a nonempty compact subset of H, for sufficiently large $w \in W$ and if $t \notin C_w$, $f(h_w(t)) = 0$. Therefore, by $(K_w.1)$,

$$(T_w f)(s) = \int_{C_w} K_w(s - h_w(t), f(h_w(t)))\, d\mu_H(t), \qquad s \in G.$$

By conditions $(K_w.1)$ and $(K_w.2)$, we have that

$$|K_w(s - h_w(t), f(h_w(t)))| \leq L_w(s - h_w(t))\psi(|f(h_w(t))|).$$

Now, by $(L_w.1)$, denoting by U a compact neighbourhood of $\theta \in G$ such that $\operatorname{supp} L_w \subset U$ for sufficiently large $w \in W$, we may deduce that $\operatorname{supp} K_w(\cdot, u) \subset U$ for sufficiently large $w \in W$ and for every $u \in \mathbb{R}$. Put now $B = U + C$. If $s \notin B$, for every $t \in C_w$, $s - h_w(t) \notin U$, and $K_w(s - h_w(t), u) = 0$ for sufficiently large $w \in W$; hence $T_w f$ vanishes outside the compact set B. Thus $\rho[\lambda(T_w f - f)] = \rho[\lambda(T_w f - f)\chi_B]$. By the (L_w, ψ)-Lipschitz condition we deduce

$$\begin{aligned}
|(T_w f)(s) - f(s)| &\leq |(T_w f)(s)| + |f(s)| \\
&= \left| \int_H K_w(s - h_w(t), f(h_w(t)))\, d\mu_H(t) \right| + |f(s)| \\
&\leq \int_H L_w(s - h_w(t))\psi(|f(h_w(t))|)\, d\mu_H(t) + |f(s)| \\
&\leq N\psi(\|f\|_\infty) + \|f\|_\infty,
\end{aligned}$$

for $s \in G$ and $w \in W$. So $\lambda|T_w f(s) - f(s)|\chi_B \leq \lambda\chi_B[N\psi(\|f\|_\infty) + \|f\|_\infty]$, for $s \in G$. Since the modular is strongly finite, we have that $\chi_B \in E_\rho$ and therefore $\rho[\lambda(T_w f - f)\chi_B] \leq \rho[\lambda(N\psi(\|f\|_\infty) + \|f\|_\infty)\chi_B] < +\infty$. By Proposition 9.2, Theorem 8.5 holds with the almost everywhere convergence of $r_n^w(s)$ (assumption K_w.3)) obtaining that $((T_w f)(s) - f(s)) \to 0$ as $w \to +\infty$, a.e. on G. Hence, applying the Lebesgue dominated convergence theorem for modular spaces (see Section 2.1), we obtain that

$$\lim_{w\to+\infty} \rho[\lambda(T_w f - f)] = 0. \qquad \square$$

Let us denote by η_H a modular on $L^0(H)$. Given $E > 0$, let

$$\mathcal{L}_E \equiv \mathcal{L}(G, H, \{h_w\}, \{\gamma_w\}, E)$$

be the subset of $L^0_{\eta_G}(G)$ whose elements f satisfy the assumption

$$\limsup_{w \to +\infty} \gamma_w \eta_H[\lambda(f \circ h_w)] \leq E \eta_G[\lambda f] \tag{9.2}$$

for every $\lambda > 0$. Then we may establish the following

Theorem 9.3. *Let $\{K_w\}_{w \in W} \subset \mathcal{K}_w$, $\{L_w\}_{w \in W} \subset \mathcal{L}^*_w$ and let ρ_G be a monotone modular and suppose that ρ_H and η_H are modulars such that the triple $\{\rho_H, \psi, \eta_H\}$ is properly directed; we assume moreover that the family $\widetilde{\mu}^s_w(A) = \int_A L_w(s - h_w(t))\, d\mu_H(t)$, $A \in \mathcal{B}(H)$, $s \in G$ is compatible with the couple (ρ_G, ρ_H). Then, given $f, g \in \operatorname{Dom} \mathbb{T}$ with $f - g \in \mathcal{L}_E$, there is an absolute constant $P > 0$, depending on $E > 0$, such that for any $\lambda \in]0, 1[$ there exists a constant $c > 0$ for which there holds*

$$\limsup_{w \to +\infty} \rho_G[c(T_w f - T_w g)] \leq P \eta_G[\lambda(f - g)].$$

Proof. Let $\lambda > 0$ be fixed with $\lambda < 1$. By the (L_w, ψ)-Lipschitz condition, we have

$$\begin{aligned}
&\rho_G[c(T_w f - T_w g)] \\
&\quad = \rho_G[c \int_H [K_w(\cdot - h_w(t), f(h_w(t))) - K_w(\cdot - h_w(t), g(h_w(t)))]\, d\mu_H(t)] \\
&\quad \leq \rho_G[c \int_H |K_w(\cdot - h_w(t), f(h_w(t))) - K_w(\cdot - h_w(t), g(h_w(t)))|\, d\mu_H(t)] \\
&\quad \leq \rho_G[c \int_H L_w(\cdot - h_w(t)) \psi(|f(h_w(t)) - g(h_w(t))|)\, d\mu_H(t)].
\end{aligned}$$

By the regularity of $\{\widetilde{\mu}^s_w\}$ defined above, with $\eta_w = \gamma_w = ||L_w||_{L^1(G)}$, and by the compatibility with the couple (ρ_G, ρ_H), we have, for sufficiently large $w \in W$,

$$\rho_G[c(T_w f - T_w g)] \leq M \gamma_w \rho_H[c D \psi(|(f - g) \circ h_w|)] + b_w.$$

Then, since $\{\rho_H, \psi, \eta_H\}$ is properly directed, we obtain for $cD \leq C_\lambda$ that

$$\begin{aligned}
\rho_G[c(T_w f - T_w g)] &\leq M \gamma_w \rho_H[C_\lambda \psi(|(f - g) \circ h_w|)] + b_w \\
&\leq M \gamma_w \eta_H[\lambda((f - g) \circ h_w)] + b_w.
\end{aligned}$$

Since $f - g \in \mathcal{L}_E$, we have

$$\limsup_{w \to +\infty} \rho_G[c(T_w f - T_w g)] \leq M E \eta_G[\lambda(f - g)].$$

The assertion easily follows by putting $P = ME$. □

Now we are ready to prove the main theorem of this section.

Theorem 9.4. *Let* $\{K_w\}_{w\in W} \subset \mathcal{K}_w$, $\{L_w\}_{w\in W} \subset \mathcal{L}_w^*$ *and let* ρ_G, η_G *be monotone, absolutely and strongly finite and absolutely continuous modulars and* ρ_H *be a modular, such that* $\{\rho_H, \psi, \eta_H\}$ *is properly directed. Let us suppose that the family* $\widetilde{\mu}_w^s(A) = \int_A L_w(s - h_w(t))\, d\mu_H(t)$, $A \in \mathcal{B}(H)$, $s \in G$, $w \in W$ *is compatible with the couple* (ρ_G, ρ_H). *Then for every* $f \in L^0_{\rho_G+\eta_G}(G)$ *such that* $f - C_c(G) \subset \mathcal{L}_E$, *there is a constant* $c > 0$ *such that*

$$\lim_{w\to+\infty} \rho_G[c(T_w f - f)] = 0.$$

Proof. Let $f \in L^0_{\rho_G+\eta_G}(G)$ be such that $f - C_c(G) \subset \mathcal{L}_E$. By Theorem 9.1, there is a $\overline{\lambda} > 0$ (we may take $\overline{\lambda} < 1$) and a sequence $\{f_n\} \subset C_c(G)$ such that

$$(\rho_G + \eta_G)[\overline{\lambda}(f_n - f)] \to 0, \quad n \to +\infty.$$

Let $\varepsilon > 0$ be fixed and let $\widetilde{n}$ be an integer such that for every $n \geq \widetilde{n}$

$$(\rho_G + \eta_G)[\overline{\lambda}(f_n - f)] < \varepsilon. \tag{9.3}$$

Fix now $\widetilde{n}$; in correspondence to such $\overline{\lambda}$ we choose a constant $c > 0$ such that $c \leq \min\left\{\frac{C_{\overline{\lambda}}}{3D}, \frac{\overline{\lambda}}{3}\right\}$. Then we have

$$\begin{aligned}\rho_G[c(T_w f - f)] &\leq \rho_G[3c(T_w f - T_w f_{\widetilde{n}})] + \rho_G[3c(T_w f_{\widetilde{n}} - f_{\widetilde{n}})] + \rho_G[3c(f_{\widetilde{n}} - f)] \\ &= I_1 + I_2 + I_3.\end{aligned}$$

Applying Theorem 9.3 to I_1, we obtain

$$\limsup_{w\to+\infty} \rho_G[3c(T_w f - T_w f_{\widetilde{n}})] \leq P\eta_G[\overline{\lambda}(f - f_{\widetilde{n}})],$$

where, without loss of generality, we can suppose $P > 1$.

Since by Theorem 9.2, we have $\lim_{w\to+\infty} \rho_G[3c(T_w f_{\widetilde{n}} - f_{\widetilde{n}})] = 0$, applying (9.3) we have

$$\limsup_{w\to+\infty} \rho_G[c(T_w f - f)] \leq P(\rho_G + \eta_G)(\overline{\lambda}(f - f_{\widetilde{n}})) \leq P\varepsilon;$$

hence the assertion follows, $\varepsilon > 0$ being arbitrary. □

9.2 Applications

In order to give some applications, for the sake of simplicity, we consider the particular case of Musielak–Orlicz spaces, where we take $H \subset G$ (H is a subgroup of G) and μ_H will denote the Haar measure of the subgroup H.

As before $\varphi : G \times \mathbb{R}_0^+ \to \mathbb{R}_0^+$ is a measurable function with respect to $s \in G$, for every $u \in \mathbb{R}_0^+$, such that $\varphi(s, \cdot)$ is a convex φ-function which satisfies inequality (9.1).

We will denote the class of functions satisfying the previous properties by $\widetilde{\Phi}$. In absence of convexity, we will denote such class simply by Φ.

Here we consider some examples of operators $T_w f$ to which the theory developed can be applied.

(I) Let $G = H = (\mathbb{R}^N, +)$ and $\mu_G = \mu_H = dt$ the Lebesgue measure. Let $I_\varphi^G(f) := I_\varphi^H(f) := \rho_G(f) = \rho_H(f) = \int_{\mathbb{R}^N} \varphi(t, |f(t)|)\, dt$, $I_\xi^G(f) = I_\xi^H(f) := \eta_G(f) = \eta_H(f) = \int_{\mathbb{R}^N} \xi(t, |f(t)|)\, dt$ with $\varphi \in \widetilde{\Phi}$, $\xi \in \Phi$, and let $L^\varphi(\mathbb{R}^N)$, $L^\xi(\mathbb{R}^N)$ be the Musielak–Orlicz spaces generated by the modulars ρ and η. Put $h_w(t) = t$ for every $w \in W$. Then, as in Section 8.3, we obtain

$$(T_w f)(s) = \int_{\mathbb{R}^N} K_w(s - t, f(t))\, dt, \quad s \in \mathbb{R}^N.$$

In this case it is clear that $\mathcal{L}_E = L^\xi(\mathbb{R}^N)$ with $E = \sup_w \gamma_w < +\infty$. Moreover, it is easy to see that in the linear case, i.e. when $K_w(s, u) = \widetilde{K}_w(s)u$, the assumptions $(K_w.i)$ and $(L_w.i)$, for $i = 1, 2, 3$, become the classical ones for approximate identities with compact support. So the theory developed includes, as particular cases, the classical convergence theorems in Musielak–Orlicz spaces, in Orlicz spaces and in L^p-spaces with $p \geq 1$ for linear integral operators of convolution type.

(II) Analogous applications can be deduced for the Mellin convolution operators of the form

$$(T_w f)(s) \equiv (\mathcal{M}_w f)(s) = \int_0^{+\infty} K_w(st^{-1}, f(t))t^{-1}\, dt, \quad s > 0$$

(see Section 8.3), where $\{K_w\}_{w \in W}$ is a suitable family of kernels. Here we take $G = H = (\mathbb{R}^+, \cdot)$, $\mu_G = \mu_H = \int t^{-1}\, dt$, $h_w(t) = t$ for every $w \in W$, $I_\varphi^G(f) = I_\varphi^H(f) := \rho_G(f) = \rho_H(f) = \int_0^{+\infty} \varphi(t, |f(t)|)t^{-1} dt$, $I_\xi^G(f) = I_\xi^H(f) := \eta_G(f) = \eta_H(f) = \int_0^{+\infty} \xi(t, |f(t)|)t^{-1} dt$ with $\varphi \in \widetilde{\Phi}$, $\xi \in \Phi$ and $L^\varphi(\mathbb{R}^+)$, $L^\xi(\mathbb{R}^+)$ are the Musielak–Orlicz spaces, respectively. Also in this case we have $\mathcal{L}_E \equiv L^\xi(\mathbb{R}^+)$ with the same E as before. The above operators are connected with the theory of moment type operators, as well as with the theory of Mellin transform. Moreover, in the case of $G = H = (\mathbb{R}_+^n, \bullet)$ with $\mathbb{R}_+^n = (]0, +\infty[)^n$ and the inner operation "$\bullet$" defined for $s = (s_1, \dots, s_n) \in \mathbb{R}_+^n$ and $t = (t_1, \dots, t_n) \in \mathbb{R}_+^n$ as $s \bullet t = (s_1 t_1, \dots, s_n t_n) \in \mathbb{R}_+^n$, the previous theory also includes the multidimensional version of the nonlinear Mellin convolution operators. In fact, G is a locally compact, topological, abelian group with neutral element $\theta = 1 = (1, \dots, 1)$, the inverse of t is given by $t^{-1} = (t_1^{-1}, \dots, t_n^{-1})$ and if we put $\langle t \rangle = \prod_{k=1}^n t_k$, the Haar measure μ_G is given by

$$\mu_G = \mu_H = \int \frac{dt}{\langle t \rangle},$$

dt being the Lebesgue measure.

(III) As in Section 8.3, let $W = \mathbb{R}^+$, $G = (\mathbb{R}, +)$, $H = (\mathbb{Z}, +)$ with μ_G the Lebesgue measure on $\mathbb{R}$, μ_H the counting measure on $\mathbb{Z}$, $h_w : \mathbb{Z} \to \mathbb{R}$ of the form $h_w(k) = \frac{k}{w}$, $k \in \mathbb{Z}$, $w > 0$. In this case we obtain the *nonlinear* version of the *generalized sampling operators* of the form

$$(T_w f)(s) = \sum_{k=-\infty}^{+\infty} K\left(ws - k, f\left(\frac{k}{w}\right)\right), \quad s \in \mathbb{R},\ w > 0,$$

where $K : \mathbb{R} \times \mathbb{R} \to \mathbb{R}$ is a kernel function with a Lipschitz condition of the form

$$|K(s, u) - K(s, v)| \leq L(s)\psi(|u - v|),$$

for every $s \in \mathbb{R}$, $u, v \in \mathbb{R}$ and for a fixed $\psi \in \Psi$, and where $L \in L^1(\mathbb{R})$ is a function with compact support.

Moreover, for $f \in L^0(\mathbb{R})$ and $g : \mathbb{Z} \to \mathbb{R}$

$$I_\varphi^G(f) := \rho_G(f) = \int_{\mathbb{R}} \varphi(s, |f(s)|)\, ds,\ I_\varphi^H(g) := \rho_H(g) = \sum_{k=-\infty}^{+\infty} \varphi(k, |g(k)|),$$

$$I_\xi^G(f) := \eta_G(f) = \int_{\mathbb{R}} \xi(s, |f(s)|)\, ds,\ I_\xi^H(g) := \eta_H(g) = \sum_{k=-\infty}^{+\infty} \xi(k, |g(k)|),$$

with functions $\varphi \in \widetilde{\Phi}$ and $\xi \in \Phi$, generating the Musielak–Orlicz spaces $L^\varphi(\mathbb{R})$, $L^\varphi(\mathbb{Z})$ and $L^\xi(\mathbb{R})$, $L^\xi(\mathbb{Z})$, respectively.

Here assumption $(K_w.3)$ becomes assumption (c) of Section 8.3 with the almost everywhere convergence instead of the uniform one and assumption $(L_w.1)$ is always satisfied since $L_w(z) := L(wz)$ has compact support and

$$\int_{\mathbb{R}} L_w(z)\, dz = \int_{\mathbb{R}} L(wz)\, dz = \frac{||L||_{L^1(\mathbb{R})}}{w} = \gamma_w.$$

Moreover, assumption $(L_w.3)$ is easily deduced in this case.

In the particular case in which K_w is linear, i.e., when $K_w(s, u) = \widetilde{K}_w(s)u$, then $L_w(s) \equiv |\widetilde{K}_w(s)|$. If in this case we suppose that

i) $\sum_{k\in\mathbb{Z}} \widetilde{K}(s - k) = 1$, for every $s \in \mathbb{R}$,

ii) $\sup_{s\in\mathbb{R}} \sum_{k\in\mathbb{Z}} L(s - k) < +\infty$,

then (c) and (b) of Section 8.3 are satisfied. The assumptions i) and ii) are very common in the theory of sampling series.

The condition (9.2) of the class $\mathcal{L}_E$ now takes on the form

$$\limsup_{w\to+\infty} \frac{1}{w} \sum_{k=-\infty}^{+\infty} \xi\left(k, \lambda\left|f\left(\frac{k}{w}\right)\right|\right) \leq S \int_{\mathbb{R}} \xi(s, \lambda|f(s)|)\, ds$$

for every $\lambda > 0$ and for some constant $S > 0$. Now, taking into account that the density result (Theorem 9.1) can be restated in the present framework with $C_c^\infty(\mathbb{R})$ instead of $C_c(\mathbb{R})$, using a C^∞-version of Urysohn's lemma, we formulate, as an application of Theorem 9.4, the following corollary.

Corollary 9.2. *Let $\varphi \in \widetilde{\Phi}$, $\xi \in \Phi$ with φ and ξ locally integrable for every $u \in \mathbb{R}$ and let $f \in L^{\varphi+\xi}(\mathbb{R})$. We suppose that $\{\rho_H, \psi, \eta_H\}$ is properly directed and that $Q(s) := ||\Gamma(\cdot, s)||_{l^1(\mathbb{Z})}$ is uniformly bounded with respect to $s \in \mathbb{R}$, where Γ is the function of Example 9.1. Moreover, let $f : \mathbb{R} \to \mathbb{R}$ be such that the following property holds*

$$\limsup_{w \to +\infty} \frac{1}{w} \sum_{k=-\infty}^{+\infty} \xi\left(k, \lambda \left| g\left(\frac{k}{w}\right)\right|\right) \leq S \int_{\mathbb{R}} \xi(s, \lambda |g(s)|)\, ds$$

for every $g \in f - C_c^\infty(\mathbb{R})$, $\lambda > 0$ and for some constant $S > 0$. Then there exists a constant $c > 0$ such that

$$\lim_{w \to +\infty} \int_{\mathbb{R}} \varphi\left(s, c \left| \sum_{k=-\infty}^{+\infty} K\left(ws - k, f\left(\frac{k}{w}\right)\right) - f(s)\right|\right) ds = 0.$$

As an example, we consider a function $\xi : \mathbb{R} \times \mathbb{R}_0^+ \to \mathbb{R}_0^+$ of the form

$$\xi(s, u) = \widetilde{\xi}(s)\gamma(u)$$

where $\widetilde{\xi} : \mathbb{R} \to \mathbb{R}_0^+$ satisfies the following conditions:

1) $\widetilde{\xi}$ is measurable,

2) there exists a constant $a > 0$ such that $\widetilde{\xi}(s) \geq a$ for every $s \in \mathbb{R}$,

3) the sequence $\varepsilon_k = \widetilde{\xi}(k)$, $k \in \mathbb{Z}$, is bounded.

We also suppose that γ is a continuous, nondecreasing function such that $\gamma(0) = 0$, $\gamma(u) > 0$ for every $u > 0$. In this case, for any function f, we have

$$\begin{aligned} \frac{1}{w} \sum_{k=-\infty}^{+\infty} \xi(k, \lambda |f(\frac{k}{w})|) &= \frac{1}{w} \sum_{k=-\infty}^{+\infty} \widetilde{\xi}\left(\frac{k}{w}\right) \gamma\left(\lambda \left| f\left(\frac{k}{w}\right)\right|\right) \frac{\widetilde{\xi}(k)}{\widetilde{\xi}(\frac{k}{w})} \\ &\leq \frac{\Gamma}{w} \sum_{k=-\infty}^{+\infty} \xi\left(\frac{k}{w}, \lambda \left| f\left(\frac{k}{w}\right)\right|\right) \end{aligned} \qquad (9.4)$$

where $\Gamma = \frac{||\widetilde{\xi}(k)||_{l^\infty}}{a}$ is a positive constant. The last term of inequality (9.4) represents a generalized Riemann sum of $\xi(\cdot, \lambda |f(\cdot)|)$. Thus, if $f : \mathbb{R} \to \mathbb{R}$ is such that the function $\xi(\cdot, \lambda |g(\cdot)|)$ is a Riemann integrable function and of bounded variation on $\mathbb{R}$, being $g \in f - C_c^\infty(\mathbb{R})$, we have that $g \in \mathcal{L}_E$ (the proof of this fact can be deduced from Theorem 3 in [114]). Thus we have, under the same assumptions of Corollary 9.2 on φ and ξ, the following result.

Corollary 9.3. *Let $\xi : \mathbb{R} \times \mathbb{R}_0^+ \to \mathbb{R}_0^+$ be of the form $\xi(s,u) = \widetilde{\xi}(s)\gamma(u)$ satisfying the previous conditions and let $\{\rho_H, \psi, \eta_H\}$ be a properly directed triple.*

If $f : \mathbb{R} \to \mathbb{R}$ is such that $f \in L^{\varphi+\xi}(\mathbb{R})$ and for every $g \in f - C_c^\infty(\mathbb{R})$ and $\lambda > 0$, the function $\xi(\cdot, \lambda|g(\cdot)|)$ is Riemann integrable and of bounded variation on $\mathbb{R}$, then there exists a constant $c > 0$ such that

$$\lim_{w \to +\infty} \int_{\mathbb{R}} \varphi\left(s, c\left|\sum_{k=-\infty}^{+\infty} K\left(ws - k, f\left(\frac{k}{w}\right)\right) - f(s)\right|\right) ds = 0.$$

Here we show that in the particular case of an Orlicz space, the previous corollary can be formulated in terms of sufficient conditions on the function f (see [202]).

First we observe that the class $\mathcal{L}_E$ contains, in particular, the set of all functions $f : \mathbb{R} \to \mathbb{R}$ such that $h_\lambda(\cdot) = \xi(\lambda|f(\cdot)|)$ is Riemann integrable on $\mathbb{R}$ for every $\lambda > 0$ and

$$\lim_{w \to +\infty} \frac{1}{w} \sum_{k=-\infty}^{+\infty} \xi\left(\lambda\left|f\left(\frac{k}{w}\right)\right|\right) = \int_{\mathbb{R}} \xi(\lambda|f(s)|)\, ds. \tag{9.5}$$

The sums in (9.5) are generalized Riemann sums of the integral on the right-hand side of (9.5). Theorem 3 in [114] proves that a characterization of the class of functions h_λ satisfying (9.5) is given by the functions $h_\lambda(\cdot)$ Riemann integrable on $\mathbb{R}$ and of *bounded coarse variation* for every $\lambda > 0$. The concept of bounded coarse variation (see [114]), is a generalization of the classical concept of bounded variation in the sense of Jordan. Thus, if in particular $h_\lambda \in BV(\mathbb{R})$ (the set of all functions with bounded variation on $\mathbb{R}$) and if it is Riemann integrable for every $\lambda > 0$, then (9.5) holds. If $BV_\xi(\mathbb{R})$ is the set of all functions such that $\xi(\lambda|f|) \in BV(\mathbb{R})$, for every $\lambda > 0$, we finally conclude that

$$\mathcal{L}_E \supset BV_\xi(\mathbb{R}) \cap E^\xi(\mathbb{R}),$$

where $E^\xi(\mathbb{R})$ denotes the space of finite elements of the Orliz space $L^\xi(\mathbb{R})$ (see also Section 7.3). Hence, under the previous assumptions on the families of kernels, we may formulate the following

Corollary 9.4. *Let $\varphi \in \widetilde{\Phi}$, $\xi \in \Phi$ with ξ locally Lipschitz in $\mathbb{R}_0^+$ and suppose that (ρ_H, ψ, η_H) is a properly directed triple. Let $f : \mathbb{R} \to \mathbb{R}$ be an absolutely Riemann integrable function on $\mathbb{R}$, and of bounded variation on $\mathbb{R}$ such that $f \in L^\varphi(\mathbb{R})$. Then there exists a constant $c > 0$ such that*

$$\lim_{w \to +\infty} \int_{\mathbb{R}} \varphi\left(c\left|\sum_{k=-\infty}^{+\infty} K\left(ws - k, f\left(\frac{k}{w}\right)\right) - f(s)\right|\right) ds = 0.$$

Proof. Since ξ is locally Lipschitz in $\mathbb{R}_0^+$, then for every $\lambda > 0$, $\xi \circ \lambda|f|$ is also Riemann integrable on $\mathbb{R}$ for every $\lambda > 0$, which means that $f \in E^\xi(\mathbb{R})$. Indeed, since $f \in BV(\mathbb{R})$, we have that f is bounded and so, since $\xi \in \Phi$ is locally Lipschitz, we have

$$|\xi(|f(t)|) - \xi(0)| \leq K|f(t)|, \quad \text{for some } K > 0.$$

Now, since $\xi(0) = 0$ and since f is absolutely Riemann integrable on $\mathbb{R}$, we deduce that $\xi \circ \lambda|f|$ is Riemann integrable on $\mathbb{R}$ for every $\lambda > 0$. Moreover, we obtain that $\xi \circ \lambda|f|$ is of bounded variation on $\mathbb{R}$ which means that $f \in BV_\xi(\mathbb{R})$. Indeed, take an arbitrary finite sequence of real numbers. In view of the boundedness of f we have

$$\begin{aligned}\sum_{i=1}^{N} |\xi(|f(t_i)|) - \xi(|f(t_{i-1})|)| &\leq K \sum_{i=1}^{N} ||f(t_i)| - |f(t_{i-1})|| \\ &\leq K \sum_{i=1}^{N} |f(t_i) - f(t_{i-1})| \leq K V_{\mathbb{R}}(f) < +\infty\end{aligned}$$

for some $K > 0$, where $V_{\mathbb{R}}(f)$ is the total variation of f in $\mathbb{R}$. Now, passing to the supremum over all the sequences of real numbers, we obtain that $(\xi \circ |f|) \in BV(\mathbb{R})$ and hence also $(\xi \circ \lambda|f|) \in BV(\mathbb{R})$. Finally we have that $f \in BV_\xi(\mathbb{R}) \cap E^\xi(\mathbb{R}) \subset \mathcal{L}_E \subset L^\xi(\mathbb{R})$. Obviously, $g \in C_c^\infty(\mathbb{R})$, is absolutely Riemann integrable and bounded on $\mathbb{R}$ and so $\lambda(f - g)$ is also absolutely Riemann integrable on $\mathbb{R}$, for every $\lambda > 0$ and bounded on $\mathbb{R}$. So, $(f - g) \in E^\xi(\mathbb{R})$, being ξ locally Lipschitz. Moreover, for $g \in C_c^\infty(\mathbb{R})$, one has $g \in BV(\mathbb{R})$, and hence $\lambda(f - g) \in BV(\mathbb{R})$ for every $\lambda > 0$ which implies that $(f - g) \in BV_\xi(\mathbb{R})$ as before. Finally, we have that $f - g \in \mathcal{L}_E$, and hence by Corollary 9.2 as formulated for Orlicz spaces, the assertion follows. □

Remark 9.1. (a) According to Remark 8.5 (a), the previous theory also contains the case of the *multivariate* sampling series of a function $f : \mathbb{R}^N \to \mathbb{R}$, in its nonlinear form. Indeed it suffices to take $G = (\mathbb{R}^N, +)$, $H = (\mathbb{Z}^N, +)$ with the Lebesgue measure and the counting measure, respectively. Hence the above generalized sampling operators take on the form

$$(T_w f)(s) = \sum_{k \in \mathbb{Z}^N} K\left(ws - k, f\left(\frac{k}{w}\right)\right),$$

for $s \in \mathbb{R}^N$, $w \in \mathbb{R}^N_+$, $K : \mathbb{R}^N \times \mathbb{R} \to \mathbb{R}$ and $f : \mathbb{R}^N \to \mathbb{R}$. Here w is a vector, i.e., $w = (w_1, \dots, w_n) \in \mathbb{R}^N_+$, and we define $w_1 \leq w_2$ if and only if $w_1^i \leq w_2^i$ for $i = 1, 2, \dots N$. Moreover if $w = (w_1, \dots, w_N)$, $s = (s_1, \dots, s_N)$, $k = (k_1, \dots, k_N)$, we set $ws = (w_1 s_1, \dots, w_N s_N)$, $\frac{k}{w} = \left(\frac{k_1}{w_1}, \dots, \frac{k_N}{w_N}\right)$ and $w \to +\infty$ means now that $w_i \to +\infty$ for each $i = 1, \dots, N$. Moreover, we have

$$\int_{\mathbb{R}^N} L_w(z)\,dz = \int_{\mathbb{R}^N} L(wz)\,dz = \frac{1}{\prod_{k=1}^N w_k} \int_{\mathbb{R}^N} L(z)\,dz = \frac{\|L\|_{L^1(\mathbb{R})}}{\prod_{k=1}^N w_k} = \gamma_w,$$

and assumption (9.2) of the class $\mathcal{L}_E$ becomes

$$\limsup_{w \to +\infty} \frac{1}{\prod_{k=1}^{N} w_k} \sum_{k \in \mathbb{Z}^N} \xi(k, \lambda |f(\frac{k}{w}|) \leq S \int_{\mathbb{R}^N} \xi(s, \lambda |f(s)|)\, ds$$

for every $\lambda > 0$ and for some constant $S > 0$.

(b) In the previous theory we may replace the real parameter $w > 0$ by an abstract parameter w varying in an arbitrary filtering partially ordered set $\mathcal{W}$.

(c) In case of $\varphi(u) = u^p$, $p \geq 1$, the previous Corollary 9.4 gives convergence results in L^p-spaces for the nonlinear sampling series of f.

(d) We remark here that, under suitable assumptions on the family of homeomorphisms $\{h_w\}$, the previous theory could also contain the case of nonuniform or irregular sampling operators in its nonlinear form.

(e) We point out that in the theory developed, we may take a more general H. Indeed, we may consider $(H, \mathcal{B}(H), \mu_H)$ as a locally compact Hausdorff topological space equipped with its Borel σ-algebra $\mathcal{B}(H)$ and with a regular measure μ_H.

9.3 Modular regular methods of summability

In order to obtain a result for the general family $(T_w)_{w \in W}$ concerning methods of regular summability in modular spaces, we give the following definition.

Given the modulars ρ_G, η_G and η_H, we say that $(T_w)_{w \in W}$ defines a *modular regular method of summability* with respect to ρ_G and η_H, if $\eta_H[\lambda(f_w - f) \circ h_w] \to 0$ as $w \to +\infty$ for some $\lambda > 0$ implies that $\rho_G[c(T_w f_w - f)] \to 0$ as $w \to +\infty$ for some $c > 0$, where $f \in L^0_{\rho_G + \eta_G}(G)$ and $f_w \in L^0(G)$.

We remark that in the case of $G = H$ and $\rho_G = \eta_H$ we obtain the definition of a regular method of summability in the modular space L^0_ρ, given in Section 5.2; indeed in this case $h_w(t) = t$, for every $w \in W$.

Now we may formulate the following

Theorem 9.5. *Under the assumptions of Theorem* 9.4, *if* $\{\gamma_w\}_{w \in W}$ *is a bounded net, we have that* $(T_w)_{w \in W}$ *defines a modular regular method of summability with respect to* ρ_G *and* η_H.

Proof. Let f_w be a family of functions in $L^0(G)$, $f \in L^0_{\rho_G + \eta_G}(G)$ and let $\lambda > 0$ be such that $\eta_H(\lambda(f_w - f) \circ h_w) \to 0$ as $w \to +\infty$. We may take $\beta \leq \min\left\{\frac{c}{2}, \frac{C_\lambda}{2D}\right\}$, c being the constant of Theorem 9.4 and C_λ the constant of the properly directed triple.

As in Theorem 8.7 of Section 8.6, we may write

$$\begin{aligned}|(T_w f_w)(s) - f(s)| &\le |(T_w f_w)(s) - (T_w f)(s)| + |(T_w f)(s) - f(s)| \\ &= I_1^w(s) + I_2^w(s).\end{aligned}$$

From the properties of the modular, we have

$$\rho_G[\beta(T_w f_w - f)] \le \rho_G[2\beta I_1^w(\cdot)] + \rho_G[2\beta I_2^w(\cdot)] = J_1(w) + J_2(w).$$

So

$$J_2(w) = \rho_G[2\beta(T_w f - f)]$$

and hence $J_2(w) \to 0$ as $w \to +\infty$ from Theorem 9.4, since $2\beta \le c$.

To evaluate $J_1(w)$ by the (L_w, ψ)-Lipschitz condition and monotonicity of ρ_G, we have

$$J_1(w) \le \rho_G[2\beta \int_H L_w(\cdot - h_w(t))\psi(|f_w(h_w(t)) - f(h_w(t))|)\, d\mu_H(t)].$$

Using the regularity of the family $\widetilde{\mu}_s^w(E) = \int_E L_w(s - h_w(t))\, d\mu_H(t)$, the compatibility of the couple (ρ_G, ρ_H), with $\eta_w = \gamma_w$, and the properly directed triple, we deduce that

$$\begin{aligned}J_1(w) &\le M\gamma_w \rho_H(2\beta D\psi(|f_w(h_w(\cdot)) - f(h_w(\cdot))|) + b_w \\ &\le M\gamma_w \eta_H(\lambda(f_w - f)\circ h_w) + b_w,\end{aligned}$$

M being the constant of the definition of compatibility for a regular net. Since $\eta_H(\lambda(f_w - f)\circ h_w) \to 0$, as $w \to +\infty$ for some $\lambda > 0$, $\{\gamma_w\}$ is a bounded net and $b_w \to 0$ as $w \to +\infty$, then $J_1(w) \to 0$ as $w \to +\infty$. Therefore the assertion follows. □

9.4 Bibliographical notes

The study of the modular convergence for the generalized sampling operators is very recent and was began by C. Bardaro and G. Vinti in [36] in Orlicz spaces. Here the modular convergence was obtained for the linear form of the general family of integral operators defined in Section 8.2, which contains, in particular, the generalized sampling series. Moreover the nonlinear case has been studied in the modular sense by G. Vinti in [202] in Orlicz spaces and then extended in [1] to Musielak–Orlicz spaces and finally to general modular spaces by I. Mantellini and G. Vinti in [144]. All these results share aspects of the approach of Butzer and his school, in the sense that an approximation process is needed in order to reconstruct the signal. The motivation for the use of the modular convergence instead of the uniform one and of the meaning of a nonlinear process in a sampling frame is pointed out in the introduction to this chapter and in Section 9.1. Moreover in [41], C. Bardaro and G. Vinti proved that the convergence process for the family of nonlinear integral operators considered in Section 8.2, defines a regular method of summability in the modular sense. For the concept of modular regular method of summability we invite the reader to see the bibliographical notes of Chapter 6.

References

[1] L. Angeloni, G. Vinti, A unified approach to approximation results with applications to nonlinear sampling theory, submitted, 2003.

[2] J. Appell, E. De Pascale, H. T. Nguyen, P. P. Zabreiko, Nonlinear integral inclusions of Hammerstein type. Contributions dedicated to Ky Fan on the occasion of his 80th birthday, *Topol. Methods Nonlinear Anal.* **5** (1995), 111–124.

[3] J. Appell, M. Dörfner, Some spectral theory for nonlinear operators, *Nonlinear Anal.* **28** (1997), 1955–1976.

[4] J. Appell, A. Vignoli, P. P. Zabreiko, Implicit function theorems and nonlinear integral equations, *Exposition. Math.* **14** (1996), 385–424.

[5] E. Baiada, La variazione totale, la lunghezza di una curva e l'integrale del Calcolo delle Variazioni in una variabile, *Atti Accad. Naz. Lincei Rend. Cl. Sci. Fis. Mat. Natur.* (8) **22** (1957), 584–588.

[6] E. Baiada, G. Cardamone, La variazione totale e la lunghezza di una curva, *Ann. Scuola Norm. Sup. Pisa Cl. Sci.* (3)**11** (1957), 29–71.

[7] E. Baiada, C. Vinti, Generalizzazioni non markoviane della definizione di perimetro, *Ann. Mat. Pura Appl.* (4) **62** (1963), 1–58.

[8] F. Barbieri, Approssimazione mediante nuclei momento, *Atti Sem. Mat. Fis. Univ. Modena* **32** (1983), 308–328.

[9] C. Bardaro, On Approximation Properties for Some Classes of Linear Operators of Convolution Type, *Atti Sem. Mat. Fis. Univ. Modena* **33** (1984), 329–356.

[10] C. Bardaro, Indipendenza dal peso della convergenza di funzionali sublineari su misure vettoriali ed applicazioni, *Rend. Istit. Mat. Univ. Trieste* **19** (1987), 44–63.

[11] C. Bardaro, D. Candeloro, Sull'approssimazione dell'integrale di Burkill–Cesari di funzionali sublineari su misure ed applicazioni all'integrale multiplo del Calcolo delle Variazioni, *Atti Sem. Mat. Fis. Univ. Modena* **26** (1977), 339–362.

[12] C. Bardaro, D. Candeloro, Teoremi di approssimazione per l'integrale multiplo del Calcolo delle Variazioni, *Rend. Circ. Mat. Palermo* (2) **30** (1981), 63–82.

[13] C. Bardaro, I. Mantellini, A modular convergence theorem for general nonlinear integral operators, *Comment. Math. Prace Mat.* **36** (1996), 27–37.

[14] C. Bardaro, I. Mantellini, Modular approximation by sequences of nonlinear integral operators in Musielak–Orlicz spaces, *Atti Sem. Mat. Fis. Univ. Modena* **46** (1998), suppl., 403–425, special volume dedicated to Prof. Calogero Vinti.

[15] C. Bardaro, I. Mantellini, Linear integral operators with homogeneous kernel: approximation properties in modular spaces. Applications to Mellin-type convolution operators and to some classes of fractional operators. In *Applied Mathematics Reviews*, Vol. 1, G. A. Anastassiou (Ed.), pp. 45–67, World Sci. Publishing, River Edge, NJ, 2000. Ed.

[16] C. Bardaro, I. Mantellini, On approximation properties of Urysohn integral operators, *Int. J. Pure Appl. Math.* 3 (2002), 129–148.

[17] C. Bardaro, J. Musielak, G. Vinti, Modular estimates and modular convergence for a class of nonlinear operators, *Math. Japon.* **39** (1994), 7–14.

[18] C. Bardaro, J. Musielak, G. Vinti, On absolute continuity of a modular connected with strong summability, *Comment. Math. Prace Mat.* **34** (1994), 21–33.

[19] C. Bardaro, J. Musielak, G. Vinti, Approximation by nonlinear integral operators in some modular function spaces, *Ann. Polon. Math.* **63** (2) (1996), 173–182.

[20] C. Bardaro, J. Musielak, G. Vinti, On the definition and properties of a general modulus of continuity in some functional spaces, *Math. Japon.* **43** (1996), 445–450.

[21] C. Bardaro, J. Musielak, G. Vinti, Some modular inequalities related to Fubini–Tonelli theorem, *Proc. A. Razmadze Math. Inst.* **119** (1998), 3–19.

[22] C. Bardaro, J. Musielak, G. Vinti, Nonlinear operators of integral type in some function spaces. Fourth International Conference on Function Spaces (Zielona Góra, 1995) *Collect. Math.* **48** (1997), 409–422.

[23] C. Bardaro, J. Musielak, G. Vinti, On nonlinear integro-differential operators in generalized Orlicz–Sobolev spaces, *J. Approx. Theory* **105** (2000), 238–251.

[24] C. Bardaro, J. Musielak, G. Vinti, On nonlinear integral equations in some function spaces, *Demonstratio Math.* **35** (2002), 583–592.

[25] C. Bardaro, S. Sciamannini, G. Vinti, Convergence in BV_φ by nonlinear Mellin-type convolution operators, *Funct. Approx. Comment. Math.* **29** (2001), 17–28.

[26] C. Bardaro, G. Vinti, Perimetro e variazione generalizzata rispetto ad una misura in $\mathbb{R}^2$, *Atti Sem. Mat. Fis. Univ. Modena* **35** (1987), 173–190.

[27] C. Bardaro, G. Vinti, Modular convergence in generalized Orlicz spaces for moment type operators, *Appl. Anal.* **32** (1989), 265–276.

[28] C. Bardaro, G. Vinti, On approximation properties of certain non convolution integral operators, *J. Approx. Theory* **62** (1990), 358–371.

[29] C. Bardaro, G. Vinti, Modular estimates of integral operators with homogeneous kernels in Orlicz type classes, *Results Math.* **19** (1991), 46–53.

[30] C. Bardaro, G. Vinti, On convergence of moment operators with respect to φ-variation, *Appl. Anal.* **41** (1991), 247–256.

[31] C. Bardaro, G. Vinti, Some estimates of certain integral operators with respect to multidimensional Vitali φ-variation and applications in Fractional Calculus, *Rend. Mat. Appl.* (7) **11** (1991), 405–416.

[32] C. Bardaro, G. Vinti, Some estimates of certain integral operators in generalized fractional Orlicz classes, *Numer. Funct. Anal. Optim.* **12** (1991), 443–453.

[33] C. Bardaro, G. Vinti, A general convergence theorem with respect to Cesari variation and applications, *Nonlinear Anal.* **22** (1994), 505–518.

[34] C. Bardaro, G. Vinti, Modular approximation by nonlinear integral operators on locally compact groups, *Comment. Math. Prace Mat.* **35** (1995), 25–47.

[35] C. Bardaro, G. Vinti, Modular estimates and modular convergence for linear integral operators. In *Mathematical Analysis, Wavelets, and Signal Processing*, Contemp. Math. 190, pp. 95–105, Amer. Math. Soc., Providence, RI, 1995.

[36] C. Bardaro, G. Vinti, A general approach to the convergence theorems of generalized sampling series, *Appl. Anal.* **64** (1997), 203–217.

[37] C. Bardaro, G. Vinti, A modular convergence theorem for certain nonlinear integral operators with homogeneous kernel. Fourth International Conference on Function Spaces (Zielona Góra, 1995), *Collect. Math.* **48** (1997), 393–407.

[38] C. Bardaro, G. Vinti, On the order of modular approximation for nets of integral operators in modular Lipschitz classes, *Funct. Approx. Comment. Math.* **26** (1998), 139–154, special issue dedicated to Prof. Julian Musielak.

[39] C. Bardaro, G. Vinti, Uniform convergence and rate of approximation for a nonlinear version of the generalized sampling operator, *Results Math.* **34** (1998), 224–240, special issue dedicated to Prof. P.L. Butzer.

[40] C. Bardaro, G. Vinti, Nonlinear sampling type operators: uniform and modular approximation results. In *Sampta 99*, Proc. Internat. Workshop on Sampling Theory and Appl., August 11–14, 1999, Loen, Norway, Norvegian University of Science and Technology, 209–215.

[41] C. Bardaro, G. Vinti, Nonlinear sampling type operators: approximation properties and regular methods of summability, *Nonlinear Anal. Forum* **6** (1) (2001), 15–26.

[42] C. Bardaro, G. Vinti, On some class of integral operators in modular spaces, *Far East J. Math. Sci.* 2001, Special Volume, Part II, 129–154.

[43] C. Bardaro, G. Vinti, Urysohn integral operators with homogeneous kernel: approximation properties in modular spaces, *Comment. Math. Prace Mat.* **42** (2) (2002), 145–182.

[44] M. G. Beaty, M. M. Dodson, Abstract harmonic analysis and the sampling theorem. In *Sampling theory in Fourier and signal analysis: advanced topics*, Oxford Science Publications, J. R. Higgins and R. L. Stens (Eds.), Oxford Univ. Press, Oxford 1999.

[45] S. N. Bernstein, Sur un procede de sommation des series trigonometriques, C. R. Acad Sci. Paris Sér. I Math. **191** (1930), 976–979.

[46] L. Bezuglaya, V. Katsnelson, The sampling theorem for functions with limited multi-band spectrum I, *Z. Anal. Anwendungen* **12** (1993), 511–534.

[47] R. Bojanic, O. Shisha, On the precision of uniform approximation of continuous functions by certain linear positive operators of convolution type, *J. Approx. Theory* **8** (1973), 101–113.

[48] M. Boni, Sull'approssimazione dell'Integrale multiplo del Calcolo delle Variazioni, *Atti Sem. Mat. Fis. Univ. Modena* **20** (1971), 187–211.

[49] M. Boni, Teoremi di approssimazione per funzionali sublineari su misure e applicazioni all'integrale del Calcolo delle Variazioni, *Atti Sem. Mat. Fis. Univ. Modena* **21** (1972), 237–263.

[50] J. L. Brown, Jr., On the error in reconstructing a non-bandlimited function by means of the band-pass sampling theorem, *J. Math. Anal. Appl.* **18** (1967), 75–84; Erratum, ibid. **21** (1968), 699.

[51] P. L. Butzer, Zur Frage der Saturationsklassen Singularer Integraloperatoren, *Math. Z.* **70** (1958), 93–112.

[52] P. L. Butzer, Representation and approximation of functions by general singular integrals I_a, I_b, *Indag. Math.* (*N.S.*) **22** (1960), 1–24.

[53] P. L. Butzer, Fourier transform methods in the theory of approximation, *Arch. Ration. Mech. Anal.* **5** (1960), 390–415.

[54] P. L. Butzer, The Banach–Steinhaus theorem with rates, and applications to various branches of analysis. In *General Inequalities II* (Proc. Second Internat. Conf., Oberwolfach, 1978), E. F. v. Beckenbach (Ed.), Internat. Schriftenreihe Numer. Math. 47, pp. 299–331, Birkhäuser, Basel 1980.

[55] P. L. Butzer, A survey of the Whittaker-Shannon sampling theorem and some of its extensions, *J. Math. Res. Exposition* **3** (1983), 185–212.

[56] P. L. Butzer, H. Berens, *Semi-groups of operators and approximation*, Grundlehren Math. Wiss. 145, Springer-Verlag, Berlin, Heidelberg, New York 1967.

[57] P. L. Butzer, W. Engels, S. Ries, R. L. Stens, The Shannon sampling series and the reconstruction of signals in terms of linear, quadratic and cubic splines, *SIAM J. Appl. Math.* **46** (1986), 299–323.

[58] P. L. Butzer, F. Fehér, Generalized Hardy and Hardy–Littlewood inequalities in rearrangement-invariant spaces, *Comment. Math.* **1** (1978), 41–64, special issue dedicated to Władysław Orlicz on the occasion of his seventy-fifth birthday.

[59] P. L. Butzer, A. Fisher, R. L. Stens, Generalized sampling approximation of multivariate signals: theory and applications, *Note Mat.* **10** (1990), 173–191.

[60] P. L. Butzer, G. Hinsen, Reconstruction of bounded signal from pseudo-periodic, irregularly spaced samples, *Signal Process.* **17** (1989), 1–17.

[61] P. L. Butzer, S. Jansche, Mellin transform theory and the role of its differential and integral operators. In *Transform Methods & Special Functions* (Proc. Second Internat. Workshop, Varna, August 1996), Bulgarian Acad. Sci., Sofia 1998.

[62] P. L. Butzer, S. Jansche, A direct approach to the Mellin Transform, *J. Fourier Anal. Appl.* **3** (1997), 325–375.

[63] P. L. Butzer, S. Jansche, The exponential sampling theorem of Signal Analysis, *Atti Sem. Mat. Fis. Univ. Modena* **46** (1998), suppl., 99–122, special volume dedicated to Prof. Calogero Vinti.

[64] P. L. Butzer, S. Jansche, A self-contained approach to Mellin transform analysis, for square integrable functions; Applications, *Integral Transform. Spec. Funct.* **8** (1999), 175–198.

[65] P. L. Butzer, S. Jansche, Mellin Fourier series and the classical Mellin transform, *Comput. Math. Appl.* **40** (2000), 49–62.

[66] P. L. Butzer, S. Jansche, R. L. Stens, Functional analytic methods in the solution of the fundamental theorems on best-weighted algebraic approximation. In *Approximation Theory* (Proc. 6th Southeast. Approximation Theory Conf., Memphis, TN, 1991) Lecture Notes in Pure Appl. Math. 138, pp. 151–205, Marcel Dekker, New York 1992.

[67] P. L. Butzer, R. J. Nessel, *Fourier Analysis and Approximation*, Pure Appl. Math. 40, Academic Press, New York, London 1971.

[68] P. L. Butzer, R. J. Nessel, De la Vallée Poussin's work in approximation and its influence, *Arch. Hist. Exact Sci.* **46** (1993), 67–95.

[69] P. L. Butzer, R. J. Nessel, Aspects of de la Vallée Poussin's work in approximation and its influence. In *Charles-Jean de la Vallée Poussin, Collected Works/Oeuvres Scientifique*, vol. I, pp. 3–9, P. L. Butzer, J. Mawhin, and P. Vetro (Eds.), Académie Royale de Belgique, Brussels; Circolo Matematico di Palermo, Palermo, 2000.

[70] P. L. Butzer, S. Ries, R. L. Stens, Shannon's sampling theorem, Cauchy's integral formula, and related results. In *Anniversary Volume on Approximation Theory and Functional Analysis* (Proc. Conf., Math. Res. Inst. Oberwolfach, Black Forest, July 30–August 6, 1983), P. L. Butzer, R. L. Stens, and B. Sz.-Nagy (Eds.), Internat. Schriftenreihe Numer. Math. 65, pp. 363–377, Birkhäuser, Basel 1984.

[71] P. L. Butzer, S. Ries, R. L. Stens, Approximation of continuous and discontinuous functions by generalized sampling series, *J. Approx. Theory* **50** (1987), 25–39.

[72] P. L. Butzer, K. Scherer, On the fundamental approximation theorems of D. Jackson, S. N. Bernstein and theorems of M. Zamansky and S. B. Steckin, *Aequationes Math.* **3** (1969), 170–185.

[73] P. L. Butzer, W. Splettstößer, A sampling theorem for duration-limited functions with error estimates, *Inform. and Control* **34** (1977), 55–65.

[74] P. L. Butzer, W. Splettstößer, R. L. Stens, The sampling theorem and linear prediction in signal analysis, *Jahresber. Deutsch. Math.-Verein* **90** (1988), 1–70.

[75] P. L. Butzer, R. L. Stens, The Poisson Summation Formula, Whittaker's cardinal series and approximate integration. In *Second Edmonton Conference on Approximation Theory* (Edmonton, Alta., 1982), CMS Conf. Proc. 3, pp. 19–36, Amer. Math. Soc., Providence, RI, 1983.

[76] P. L. Butzer, R. L. Stens, The Euler–McLaurin summation formula, the sampling theorem and approximate integration over the real axis, *Linear Algebra Appl.* **52/53** (1983), 141–155.

[77] P. L. Butzer, R. L. Stens, A modification of the Wittaker–Kotelnikov–Shannon sampling series, *Aequationes Math.* **28** (1985), 305–311.

[78] P. L. Butzer, R. L. Stens, Sampling theory for not necessarily band-limited functions: a historical overview, *SIAM Rev.* **34** (1992), 40–53.

[79] P. L. Butzer, R. L. Stens, Linear prediction by samples from the past. In *Advanced Topics in Shannon Sampling and Interpolation Theory*, R. J. Marks II (Ed.), Springer Texts Electrical Engrg., pp. 157–183, Springer, New York 1993.

[80] D. Candeloro, P. Pucci, L'integrale di Burkill–Cesari su un rettangolo e applicazioni all'integrale di Fubini–Tonelli relativamente a coppie di curve continue, *Boll. Un. Mat. Ital.* **17-B**, (1980) 835–859.

[81] D. Candeloro, P. Pucci, L'integrale di Burkill-Cesari come integrale del Calcolo delle Variazioni, *Boll. Un. Mat. Ital.* **18-B** (1981), 1–24.

[82] L. Cesari, Sulle funzioni a variazione limitata, *Ann. Scuola Norm. Sup. Pisa Cl. Sci* (2) **5** (1936), 299–313.

[83] L. Cesari, Quasiadditive set functions and the concept of integral over a variety, *Trans. Amer. Math. Soc.* **102** (1962), 94–113.

[84] F. Degani Cattelani, Nuclei di tipo distanza che attutiscono i salti in una o piú variabili, *Atti Sem. Mat. Fis. Univ. Modena* **30** (1981), 299–321.

[85] F. Degani Cattelani, Approssimazione del perimetro di una funzione mediante nuclei di tipo momento, *Atti Sem. Mat. Fis. Univ. Modena* **34** (1985–86), 145–168.

[86] E. De Giorgi, Definizione ed espressione analitica di perimetro di un insieme, *Atti Accad. Naz. Lincei Rend. Cl. Sci. Fis. Mat. Natur.* (7) **14** (1953), 390–393.

[87] E. De Giorgi, Su una teoria generale della misura $(r-1)$-dimensionale in uno spazio ad r dimensioni, *Ann. Mat. Pura Appl.* (4) **36** (1954), 191–213.

[88] Ch.-J. De La Vallée Poussin, Sur la convergencen des formules d'interpolation entre ordonnées équidistantes, *Bull. Cl. Sci. Acad. Roy. Belg.* **4** (1908), 319–410.

[89] W. Dickmeis, R. J. Nessel, A unified approach to certain counterexamples in approximation theory in connection with a uniform boundedness principle with rates, *J. Approx. Theory* **31** (1981), 161–174.

[90] W. Dickmeis, R. J. Nessel, A quantitative condensation of singularities on arbitrary sets, *J. Approx. Theory* **43** (1985), 383–393.

[91] M. M. Dodson, A. M. Silva, Fourier analysis and the sampling theorem, *Proc. Roy. Irish. Acad. Sect. A* **85** (1985), 81–108.

[92] T. Dominguez-Benavides, M. A. Khamsi, S. Samadi, Asymptotically regular mappings in modular function spaces, *Sci. Math. Japon.* **53** (2001), 295–304.

[93] T. Dominguez-Benavides, M. A. Khamsi, S. Samadi, Uniformly lipschitzian mappings in modular function spaces, *Nonlinear Anal.* **46** (2001), 267–278.

[94] T. Dominguez-Benavides, M. A. Khamsi, S. Samadi, Asymptotically nonexpansive mappings in modular function spaces, *J. Math. Anal. Appl.* **265** (2002), 249–263.

[95] N. Dunford, J. T. Schwartz, *Linear Operators, I. General Theory*, Pure Appl. Math. 7, Interscience Publishers, New York, London 1958.

[96] A. Erdelyi, On fractional integration and its application to the theory of Hankel transforms, *Quart. J. Math. Oxford Ser.* (2) **11** (1940), 293–303.

[97] F. Fehér, A generalized Schur–Hardy inequality on normed Köthe spaces. In *General Inequalities II* (Proc. Second Internat. Conf., Oberwolfach, 1978), E. F. v. Beckenbach (Ed.), Internat. Schriftenreihe Numer. Math. 47, pp. 277–285, Birkhäuser, Basel 1980.

[98] C. Fiocchi, Nuclei momento due-dimensionali e convergenza in area, *Atti Sem. Mat. Fis. Univ. Modena* **33** (1984), 291–312.

[99] C. Fiocchi, Variazione di ordine α e dimensione di Hausdorff degli insiemi di Cantor, *Atti Sem. Mat. Fis. Univ. Modena* **34** (1991), 649–667.

[100] T. M. Flett, A note on some inequalities, *Proc. Glasgow Math. Assoc.* **4** (1958), 7–15.

[101] G. B. Folland, *Real Analysis. Modern Techniques and their Applications*, Pure Appl. Math., A Wiley-Interscience Publication. John Wiley & Sons, Inc., New York, 1984.

[102] G. B. Folland, E. M. Stein, *Hardy spaces on homogeneous groups*, Math. Notes 28, Princeton Univ. Press, Princeton 1982.

[103] I. Fredholm, Sur une classe d'equations fonctionelle, *Acta Math.* **27** (1903), 365–390.

[104] E. Giusti, *Minimal surfaces and functions of bounded variation*, Monographs Math. 80, Birkhäuser, Basel 1984.

[105] S. Gniłka, On the generalized Helly's theorem, *Funct. Approx. Comment. Math.* **4** (1976), 109–112.

[106] S. Gniłka, On the approximation of M-absolutely continuous functions. I. Approximation by step-functions, *Funct. Approx. Comment. Math.* **4** (1976), 113–123.

[107] S. Gniłka, Remarks on the generalized absolute continuity, *Funct. Approx. Comment. Math.* **5** (1977), 39–45.

[108] S. Gniłka, On the approximation of M-absolutely continuous functions. II. Approximation by Steklov functions, *Funct. Approx. Comment. Math.* **5** (1977), 161–166.

[109] S. Gniłka, On the approximation of M-absolutely continuous functions. III. Approximation by singular integrals, *Funct. Approx. Comment. Math.* **5** (1977), 167–169.

[110] S. Gniłka, Modular spaces of functions of bounded M-variation, *Funct. Approx. Comment. Math.* **6** (1978), 3–24.

[111] C. Goffman, J. Serrin, Sublinear functions of measures and variational integrals, *Duke Math. J.* **31** (1964), 159–178.

[112] A. Gogatishvili, V. Kokilashvili, Criteria of weighted inequalities in Orlicz classes for maximal function defined on homogeneous type spaces, *Georgian Math. J.* **1** (1994), 641–673.

[113] D. Guo, V. Lakshmikantham, X. Liu, *Nonlinear Integral Equations in Abstract Spaces*, Math. Appl. 373, Kluwer Acad. Publ., Dordrecht 1996.

[114] S. Haber, O. Shisha, Improper integrals, simple integrals and numerical quadrature, *J. Approx. Theory* **11** (1974), 1–15.

[115] G. H. Hardy, J. E. Littlewood, A convergence criterion for Fourier series, *Math. Z.* **28** (1928), 612–634.

[116] H. H. Herda, Modular spaces of generalized variation, *Studia Math.* **30** (1968), 21–42.

[117] E. Hewitt, K. A. Ross, *Abstract Harmonic Analysis. I. Structure of topological groups. Integration theory, group representations*, Grundlehren Math. Wiss. 115, Springer-Verlag, Berlin 1963.

[118] J. R. Higgins, Five short stories about the cardinal series, *Bull. Amer. Math. Soc.* **12** (1985), 45–89.

[119] J. R. Higgins, *Sampling Theory in Fourier and Signal Analysis: Foundations*, Oxford Univ. Press, Oxford 1996.

[120] J. R. Higgins and R. L. Stens (Eds.), *Sampling theory in Fourier and signal analysis: advanced topics*, Oxford Science Publications, Oxford Univ. Press, Oxford 1999.

[121] T. H. Hildebrandt, *Introduction to the Theory of Integration*, Pure Appl. Math. 13, Academic Press, New York, London 1963.

[122] A. J. Jerri, The Shannon sampling-its various extensions and applications: a tutorial review, *Proc. IEEE* **65** (1977), 1565–1596.

[123] A. Kamińska, On some compactness criterion for Orlicz subspace $E_\Phi(\Omega)$, *Comment. Math. Prace Mat.* **22** (1981), 245–255.

[124] A. Kamińska, R. Płuciennik, Some theorems on compactness in generalized Orlicz spaces with application of the Δ_∞-condition, *Funct. Approx. Comment. Math.* **10** (1980), 135–146.

[125] M. A. Khamsi, Fixed point theory in modular function spaces, In *Recent Advances on Metric Fixed Point Theory* (Sevilla, 1995), T. Domínguez Benavides (Ed.), Ciencias 48, pp. 31–57, Univ. Sevilla, Sevilla 1996.

[126] M. A. Khamsi, W. M. Kozlowski, S. Reich, Fixed point theory in modular function spaces, *Nonlinear Anal.* **14** (1990), 935–953.

[127] I. Kluvánek, Sampling theory in abstract harmonic analysis, Mat.-Fyz. Časopis Sloven. Akad. Vied **15** (1965), 43–48.

[128] H. Kober, On fractional integrals and derivatives, *Quart. J. Math.* **11** (1940), 193–211.

[129] V. Kokilashvili, M. Krbec, *Weighted Inequalities in Lorentz and Orlicz Spaces*, World Scientific Publishing Co., Inc., River Edge, NJ, 1991.

[130] V. A. Kotel'nikov, On the carrying capacity of "ether" and wire in elettrocommunications. In *Material for the First All-Union Conference on Questions of Communications*, Izd. Red. Upr. Svyazi RKKA, Moscow, 1933 (in Russian); English translation in *Appl. Numer. Harmon. Anal. Modern Sampling Theory*, pp. 27–45, Birkhäuser, Boston, MA, 2001.

[131] W. M. Kozlowski, *Modular function spaces*, Pure Appl. Math. 122, Marcel Dekker, New York, Basel 1988.

[132] M. A. Krasnosel'skii, *Topological Methods in the Theory of Nonlinear Integral Equations*, Internat. Ser. Monographs Pure Appl. Math. 45, Pergamon Press, Oxford 1964.

[133] M. A. Krasnosel'skii, Y. B. Rutickii, *Convex Functions and Orlicz Spaces*, P. Noordhoff Ltd., Groningen 1961.

[134] R. Leśniewicz, W. Orlicz, On generalized variations. II, *Studia Math.* **45** (1973), 71–109.

[135] E. R. Love, A generalization of absolute continuity, *J. London Math. Soc.* **26** (1951), 1–13.

[136] E. R. Love, Some inequalities for fractional integrals. In *Linear Spaces and Approximation* (Proc. Conf., Math. Res. Inst., Oberwolfach, 1977), P. L. Butzer and B. Szökefalvi-Nagy (Eds.), Internat. Schriftenreihe Numer. Math. 40, pp. 177–184, Birkhäuser, Basel 1978.

[137] E. R. Love, Links between some generalizations of Hardy's integral inequality. In *General Inequalities IV* (Proc. Fourth Internat. Conf., Oberwolfach, 1983), W. Walter (Ed.), Internat. Ser. Numer. Math. 71, pp. 45–57, Birkhäuser, Basel 1984.

[138] E. R. Love, L. C. Young, Sur une classe de fonctionnelles linèaires, *Fund. Math.* **28** (1937), 243–257.

[139] W. A. Luxemburg, A. C. Zaanen, *Riesz spaces I*, North-Holland, Amsterdam, London 1971.

[140] L. Maligranda, *Orlicz spaces and Interpolation*, Seminários de Matemática 5, Campinas, SP: Univ. Estadual de Campinas, Dep. de Matemática, 1989.

[141] I. Mantellini, Generalized sampling operators in modular spaces, *Comment. Math. Prace Mat.* **38** (1998), 77–92.

[142] I. Mantellini, G. Vinti, Modular estimates for nonlinear integral operators and applications in fractional calculus, *Numer. Funct. Anal. Optim.* **17** (1998), 143–165.

[143] I. Mantellini, G. Vinti, Φ-variation and nonlinear integral operators, *Atti Sem. Mat. Fis. Univ. Modena* **46** (1998), suppl., 847–862, special volume dedicated to Prof. Calogero Vinti.

[144] I. Mantellini, G. Vinti, Approximation results for nonlinear integral operators in modular spaces and applications, *Ann. Polon. Math.* **81** (1) (2003), 55–71.

[145] C. Martínez, M. Sans, M. D. Martínez, Some inequalities for fractional integrals and derivatives, *Dokl. Akad. Nauk*, **315** (1990), 1049–1052; English translation in *Soviet Math. Dokl.* **42** (1991), 876–879.

[146] K. E. Michener, Weak convergence of the area of nonparametric L_1-surfaces, *Trans. Amer. Math. Soc.* **234** (1977), 175–184

[147] A. Musielak, J. Musielak, On nonlinear integral operators in function spaces, *Math. Japon.* **48** (1998), 257–266.

[148] J. Musielak, On some modular spaces connected with strong summability, *Math. Student* **27** (1959), 129–136.

[149] J. Musielak, Sequences of finite M-variation, *Prace Mat.* **6** (1961), 165–136 (in Polish).

[150] J. Musielak, An application of modular spaces to approximation, *Comment. Math.* **1** (1978), 251–259, special issue dedicated to Władysław Orlicz on the occasion of his seventy-fifth birthday.

[151] J. Musielak, Modular approximation by a filtered family of linear operators. In *Functional Analysis and Approximation* (Proc. Conf., Math. Res. Inst., Oberwolfach, August 9–16, 1980), P. L. Butzer, Béla Sz.-Nagy and E. Görlich (Eds.), Internat. Schriftenreihe Numer. Math. 60, pp. 99–110, Birkhäuser, Basel 1981.

[152] J. Musielak, On some approximation problems in modular spaces. In *Constructive Function Theory 1981* (Proc. Int. Conf., Varna, June 1–5, 1981), pp. 455–461, Publ. House Bulgarian Acad. Sci., Sofia 1983.

[153] J. Musielak, *Orlicz spaces and modular spaces*, Lecture Notes in Math. 1034, Springer-Verlag, Berlin 1983.

[154] J. Musielak, Approximation by nonlinear singular integral operators in generalized Orlicz spaces, *Comment. Math. Prace Mat.* **31** (1991), 79–88.

[155] J. Musielak, Nonlinear approximation in some modular function spaces I, *Math. Japon.* **38** (1993), 83–90.

[156] J. Musielak, On the approximation by nonlinear integral operators with generalized Lipschitz kernel over a compact abelian group, *Comment. Math. Prace Mat.* **33** (1993), 99–104.

[157] J. Musielak, Approximation by nonlinear integral operators with generalized Lipschitz kernel over a compact abelian group, *Atti Sem. Mat. Fis. Univ. Modena* **43** (1995), 225–228.

[158] J. Musielak, Nonlinear integral operators with generalized Lipschitz kernels. In *Functional Analysis* (Proc. of First Int. Workshop, Trier University, September 26–October 1, 1994), pp. 313–318, de Gruyter, Berlin 1996.

[159] J. Musielak, On some conservative nonlinear integral operators, *Acta Univ. Lodz. Folia Math.* **8** (1996), 23–31.

[160] J. Musielak, Approximation in modular function spaces, *Funct. Approx. Comment. Math.* **25** (1997), 45–57.

[161] J. Musielak, On nonlinear integral operators, *Atti Sem. Mat. Fis. Univ. Modena* **47** (1999), 183–190.

[162] J. Musielak, Approximation by a nonlinear convolution operator in modular function spaces, *Rocznik Nauk.-Dydakt. Prace Mat.* **17**, (2000), 181–190.

[163] J. Musielak ,W. Orlicz, On generalized variations. I, *Studia Math.* **18** (1959), 11–41.

[164] J. Musielak, W. Orlicz, On modular spaces, *Studia Math.* **18** (1959), 49–65.

[165] J. Musielak, W. Orlicz, On modular spaces of strongly summable sequences, *Studia Math.* **22** (1962), 127–146.

[166] H. Nakano, *Modulared Semi-Ordered Linear Spaces*, Maruzen Co., Ltd., Tokyo 1950.

[167] H. Nakano, *Topology of Linear Topological Spaces*, Maruzen Co., Ltd., Tokyo 1951.

[168] K. Ogura, On a certain transcendental integral function in the theory of interpolation, *Tôhoku Math. J.* **17** (1920), 64–72.

[169] K. B. Oldham, J. Spanier, *The Fractional Calculus*, Math. Sci. Engrg. 111, Academic Press, New York, London 1974.

[170] W. Orlicz, Über eine gewisse Klasse von Räumen vom Typus B, *Bull. Acad. Polon. Sci. Lett.*, Ser. A (1932), 207–220.

[171] W. Orlicz, Über Räume L^M, *Bull. Acad. Polon. Sci. Lett.*, Ser. A (1936), 93–107.

[172] D. Pallaschke, S. Rolewicz, *Foundations of Mathematical Optimization. Convex Analysis without linearity*, Math. Appl. 388, Kluwer Academic Publishers, Dordrecht 1997.

[173] T. Radó, *Length and Area*, Amer. Math. Soc. Colloq. Publ. 30, Amer. Math. Soc., Providence 1948.

[174] M. M. Rao, Z. D. Ren, *Theory of Orlicz Spaces*, Monographs Textbooks Pure Appl. Math. 146, Marcel Dekker, Inc., New York 1991.

[175] S. Ries, R. L. Stens, Approximation by generalized sampling series, in *Constructive Theory of Functions*, Sendov, Bl., Petrushev, P., Maleev, R., Tashev, S. (Eds.), Publishing House of the Bulgarian Academy of Sciences, Sofia, 1984, 746–756.

[176] L. L. Schumaker, *Spline Functions: Basic Theory*, John Wiley & Sons, New York 1981.

[177] S. Sciamannini, G. Vinti, Convergence and rate of approximation in BV_φ for a class of integral operators, *Approx. Theory Appl.* **17** (2001), 17–35.

[178] S. Sciamannini, G. Vinti, Convergence results in BV_φ for a class of nonlinear Volterra type integral operators and applications, to appear in *J. Concrete Appl. Anal.*, 2003.

[179] C. E. Shannon, Communication in the presence of noise, *Proc. I.R.E* **37** (1949), 10–21.

[180] P. C. Sikkema, Approximation formulae of Voronoskaja type for certain convolution operators, *J. Approx. Theory* **26** (1979), 26–45.

[181] P. C. Sikkema, Fast approximation by means of convolution operators, *Indag. Math.* **43** (1981), 431–444.

[182] S. Singh, B. Watson, P. Srivastava, *Fixed Point Theory and Best Approximation: The KKM-map Principle*, Math. Appl. 424, Kluwer Academic Publishers, Dordrecht 1997.

[183] I. Someya, *Waveform transmission*, Shukyo, Tokyo 1949.

[184] W. Splettstößer, Error estimates for sampling approximation of non-bandlimited functions, *Math. Methods Appl. Sci.* **1** (1979), 127–137.

[185] W. Splettstößer, 75 years aliasing error in the sampling theorem. In *Signal Processing II: Theories and Applications* (Proc. Second Signal Processing Conf. (EUSIPCO-83), Erlangen, September 12–16, 1983), H. W. Schüssler (Ed.), pp. 1–4, North-Holland, Amsterdam 1983.

[186] R. L. Stens, Approximation of duration-limited functions by sampling sums, *Signal Process.* **2** (1980), 173–176.

[187] R. L. Stens, Approximation of functions by Whittaker's cardinal series. In *General Inequalities IV*, (Proc. Fourth Internat. Conf., Oberwolfach, 1983), W. Walter (Ed.), Internat. Ser. Numer. Math. 71, pp. 137–149, Birkhäuser Verlag, Basel 1984.

[188] J. Szelmeczka, On convergence of singular integrals in the generalized variation metric, *Funct. Approx. Comment. Math.* **15** (1986), 53–58.

[189] M. Theis, Über eine Interpolationsformel von de la Vallée Poussin, *Math. Z.* **3** (1919), 93–113.

[190] B. Tomasz, Nonlinear integral operators and methods of summability, *Comment. Math. Prace Mat.* **36** (1996), 247–254.

[191] B. Tomasz, On consistency of summability methods generated by nonlinear integral operators, *Comment. Math. Prace Mat.* **39** (1999), 197–205.

[192] L. Tonelli, Sulla quadratura delle superfici, I e II, *Atti Accad. Naz. Lincei Rend. Cl. Sci. Fis. Mat. Nat.* (6) **3** (1926), 357–362, 633–638.

[193] M. Väth, Approximation, complete continuity and uniform measurability of Uryson operators on general measure spaces, *Nonlinear Anal.* **33** (1998), 715–728.

[194] M. Väth, A compactness criterion of mixed Krasnoselskii–Riesz type in regular ideal spaces of vector functions, *Z. Anal. Anwendungen* **18** (1999), 713–732.

[195] M. Väth, *Volterra and integral equations of vector functions*, Pure Appl. Math. 224, Marcel Dekker Inc., New York, Basel 2000.

[196] C. Vinti, Sopra una classe di funzionali che approssimano l'area di una superficie, *Ann. Mat. Pura e Appl.* (4) **48** (1959), 237–255.

[197] C. Vinti, Perimetro – Variazione, *Ann. Scuola Norm. Sup. Pisa Cl. Sci.* (3) **18** (1964), 201–231.

[198] C. Vinti, Sull'approssimazione in perimetro e in area, *Atti Sem. Mat. Fis. Univ. Modena* **13** (1964), 187–197.

[199] C. Vinti, Espressioni che danno l'area di una superficie, *Atti Sem. Mat. Fis. Univ. Modena* **17** (1968), 289–350.

[200] C. Vinti, A Survey on recent results of the Mathematical Seminar in Perugia, inspired by the Work of Professor P. L. Butzer, *Results Math.* **34** (1998), 32–55.

[201] G.Vinti, Generalized φ-variation in the sense of Vitali: estimates for integral operators and applications in fractional calculus, *Comment. Math. Prace Mat.* **34** (1994), 199–213.

[202] G.Vinti, A general approximation result for nonlinear integral operators and applications to signal processing, *Appl. Anal.* **79** (2001), 217–238.

[203] A. Waszak, Strong summability of functions in Orlicz metrics, *Commentationes Math.* **12** (1968), 115–139.

[204] P. Weiss, An estimate of the error arising from misapplication of the sampling theorem, *Notices Amer. Mat. Soc.* **10** (1963), 351.

[205] J. D. Weston, The cardinal series in Hilbert space, *Math. Proc. Cambridge Philos. Soc.* **45** (1949), 335–341.

[206] E. T. Whittaker, On the functions which are represented by the expansion of the interpolation theory, *Proc. Roy. Soc. Edinburgh* **35** (1915), 181–194.

[207] J. M. Whittaker, The "Fourier" theory of the cardinal function, *Proc. Edinburgh Math. Soc.* (2) **1** (1927–1929), 169–176.

[208] N. Wiener, The quadratic variation of a function and its Fourier coefficents, *Journ. Mass. Inst. of Technology* **3** (1924), 73–94.

[209] L. C. Young, An inequality of Hölder type, connected with Stieltjes integration, *Acta Math.* **67** (1936), 251–282.

[210] L. C. Young, General inequalities for Stieltjes integrals and the convergence of Fourier series, *Math. Ann.* **115** (1938), 581–612.

[211] P. P. Zabreiko, Ideal spaces of functions I, *Vestnik Jaroslav. Univ. Vyp.* **8** (1974), 12–52 (in Russian).

[212] V. Zanelli, Funzioni momento convergenti dal basso in variazione di ordine non intero, *Atti Sem. Mat. Fis. Univ. Modena* **30** (1981), 355–369.

Index